This journal belongs to:

_____

DR. FUN GUY'S

MUSHROOM JOURNAL

AND

FORAGING GUIDE

DR. GORDON WALKER

Clarkson Potter/Publishers
New York

CONTENTS

INTRODUCTION

Welcome to the wild, wonderful world of mushrooms and lichens! This journal is a companion to my book *Dr. Fun Guy's Passport to Kingdom Fungi*, which offers a deep dive into the basics of fungal biology. Whereas *Passport* is an introductory overview of all things fungi, this journal focuses on identification, giving you the chance to seek out mushrooms and write about your findings.

My hope is that it will help shine a light on the incredible diversity of fungi and all their different forms. Within these pages, you will find many of the most common edible, toxic, and medicinal mushrooms along with other fungi like rusts, smuts, and lichens. However, this is not an exhaustive guide, as there are hundreds more mushroom types not covered in this book.

Identifying mushrooms is a skill you can develop through curiosity, diligence, and repetition—don't be discouraged if you can't figure everything out right away. Use this book to start categorizing your finds based on the sixteen different morphological types (page 17). As you start to recognize more types of mushrooms and their scientific names (aka Latin binomials), check out the taxonomic tree by scanning the QR code on page 232. It will help you get a better sense of how different mushroom families are related without relying solely on their distinguishing visual features—or, as mycologists would call it, a mushroom's morphology.

To make this journal broadly usable regardless of where you live, mushrooms are typically presented in a general sense at the genus level and when appropriate at the species level. I provide categorized descriptions and resources to help you recognize what you find in nature, but the information in this book is not meant for exact species level identification. My intention is that you use this book to help you navigate others, especially when it comes to understanding complex terms and defined categories of mushrooms. For focusing on local species identification, I suggest you find a recently published regional guidebook that is specific to your geographic area.

How to use this journal

Each section of this journal focuses on a specific mushroom type, including a selection of mushrooms that you can seek out in nature. On these pages, you'll find information about a mushroom's taxonomy (genetic lineage), edibility/toxicity, ecological strategy, spore color, general appearance, look-alikes, and notable species. Each entry also includes photos, but given the inherent variability of mushrooms, more are often necessary to definitively identify a mushroom. To that end, I've also provided QR codes that will link you to thousands more photos on iNaturalist, the best free identification resource and community science platform. From there, you can seek out mushrooms in nature.

Once you've found your mushroom, you can journal about your findings. The mushroom entries in this book have a corresponding journal page where you can document the distinct features of the mushrooms that grow in your area. On these pages, you'll be able to take notes on a mushroom's appearance, shape, texture, and smell, along with where and how they are growing. You can use these notes to inform future trips—perhaps you want to return to a specific site, or maybe you want to avoid it the following year. You can also use this book as a memento for looking back on your adventures.

Several of this book's sections conclude with "honorable mentions" where I've listed other types of relevant mushrooms and corresponding photos to assist you with identification and recognition. These are followed by blank journaling templates where you can document the mushrooms from the honorable mentions. Or you can use that space to write about mushrooms that aren't included—the choice is yours. With the honorable mentions section, my goal is to prompt you to record details of mushrooms that you want to investigate further.

Along with my book *Passport to Kingdom Fungi*, *please* consider this journal an essential part of your road map to learning the ropes of mushroom identification.

What are the best places to find mushrooms?

The best habitats to find a broad diversity of mushrooms are forested areas with a mixture of different tree species. Healthy forest ecosystems support a wide diversity of mushrooms. The combination of trees, abundant decaying organic matter, and copious amounts of rainfall and moisture encourage mushroom growth. Mushrooms tend to fruit in microclimates within the forest, often in proximity to specific plant species. Learning to identify trees and other plants will help you focus on specific areas, greatly increasing your chances of finding the best edible mushrooms. The more you learn about the forest as a whole, the more you can use your intuition to find particularly productive mushroom patches. While forests tend to have the widest selection of mushrooms, you can also find a huge diversity of species associated with grasslands, chaparrals, deserts, mountains, and even urban habitats. Often, it's just a matter of learning how to spot moist areas that provide the perfect conditions for the formation of fruiting bodies.

What kind of weather is best for mushrooms?

The main ingredient you need for mushrooms is moisture. Whether that comes in the form of rain, fog, or melting snow depends on where you are in the world and the type of habitat you are hunting in. If you want to be a successful mushroom hunter, you need to pay attention to the weather and the seasons, particularly daily temperatures, rainfall, and relative humidity. The time of year will help inform you about what kinds of mushrooms you should be looking for. In general, the best time to hunt for mushrooms is one to three days after a heavy soaking rain; this gives new mushrooms time to grow and maturing mushrooms time to dry out. Most mushrooms tend to fruit best with moderate temperatures between 50 and 85°F and humidity above 80 percent. Most mushrooms will not grow when it's freezing or when conditions are very dry or hot, although there are specially adapted species and exceptions. Saprotrophic species tend to actively fruit during rainy wet conditions, while mycorrhizal species can emerge up to three weeks after a major rain event. Many of the large fleshy edible polypores can form fruiting bodies regardless of rain, drawing on moisture from deep in their woody substrate.

What seasons are best for finding mushrooms?

Each geographic region has its own seasonal arc for mushrooms. Some mushrooms tend to have very short seasons, like ephemeral morels (*Morchella*) popping up for a week or two in the spring. Chanterelles (*Cantharellus*) tend to have long seasonal arcs, growing for many months, while oyster mushrooms (*Pleurotus*) can fruit whenever conditions are advantageous. In general, temperature and rainfall tend to dictate when particular species will appear. Thus, the best season to find mushrooms depends on where you live and what types of habitats you have access to. For much of North America, including the Eastern Seaboard, Midwest, and Southeast, the seasonal arc for mushrooms tends to start in spring and run through the fall until heavy frost and snows return (usually from April until the end of October). This trend is similar for other temperate regions in Europe and Asia in the Northern Hemisphere. Temperate regions in the Southern Hemisphere display a similar seasonal arc for mushrooms, but at the opposite calendar time of year (generally between December and May). For the coastal areas of the Pacific Northwest and Northern California, the seasonal arc starts in late summer (with fog drip) and runs through spring, depending on rainfall and snow (generally August to February).

What are your tips for finding mushrooms?

The best way to find a mushroom is to think like a mushroom. Mushrooms tend to fruit in damp areas with well-draining soil and lots of organic material. As you walk around the woods, look for areas with more moisture, like streams, drainage channels, low-lying areas, and shaded spots. Fruiting bodies tend to form along the drip line of trees and under dense foliage. Also be sure to check under bushes, on or beside fallen logs, and among the litter of leaves and twigs called duff that covers the forest floor. Many mushrooms like disturbance, so the interface between the forest and the path is a surprisingly good place to look. When you come across a part of the path with lots of fruiting bodies, it's a good idea to selectively go off trail and explore further. While out hiking, remember to slow down and give your eyes time to adjust to small perturbations on the forest floor. Look for colors and patterns of mushrooms that are distinct from plants. Sunlight helps highlight the unique luminosity of mushrooms, making them stand out against the forest floor. Pay attention to animal activity, as dig holes and scattered bits can be a sign of mushrooms nearby. Inherently, many mushrooms are often somewhat hidden as they grow. Frequently, fruiting bodies will push up the duff, thereby forming little humps, affectionately known to foragers as shrumps. Learning to spot these shrumps will help you find many more mushrooms, especially prime edible specimens.

Best Practices for Mushroom Hunting

The long history of human foraging practices, systematic scientific studies, and anecdotal evidence all indicate that individuals' harvesting of mushrooms is sustainable behavior. Responsibility is a built-in requirement of foraging and spending time in nature. Every time you go out searching for wild foods, you are taking your life into your own hands. Eating the wrong plant or mushroom can be a deadly mistake. However, the inherent responsibilities of a forager extend beyond personal safety. If you benefit from taking food from nature, then you have a duty to protect the land that provides for you. Here are some suggestions that can help you forage ethically, meaning your activities help support and foster healthy foraging habitats.

- Look up and follow local rules and regulations for collecting mushrooms.
- Obtain permits and/or permission for the land you are foraging on when necessary.
- In general, when foraging, take only what you need and can reasonably use (or at least share with friends and family). Do not get greedy; leave some behind for animals that eat mushrooms, including other foragers.
- Use what you take; be ready to process and preserve your foraged foods.
- Carry your mushrooms in a breathable container (like a mesh bag or woven basket) that will allow them to continue dispersing spores as you move around.

- Learn to recognize invasive plant species and remove them (especially if they are edible).
- Clean and sanitize the soles of your shoes between hikes to help avoid spreading pathogens between different areas.
- When possible, stay on trails and established paths so as not to compact the forest floor.
- When off path, avoid trampling plants and causing collateral damage to the habitat.
- Help make the forest cleaner than when you came. Carry an extra bag for picking up trash, especially plastics and metals that do not biodegrade.
- Whenever possible, leave no trace. Treat natural spaces with reverence and respect.
- Lead by example, practice what you preach, and think about how your behavior can influence others.

How to Document and Voucher Mushrooms: A Photography Guide

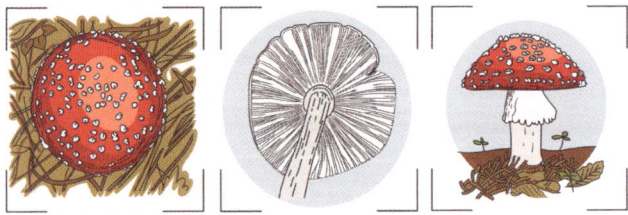

Figure 1. Take photos of the top, bottom, and side(s) of the mushroom. Capture the surrounding scene as well, including the nearby plants and the specific habitat. If possible, take a photo of the mushroom at different stages of its development. Be sure to include the natural spore print/deposit in your photos, plus an object for scale.

 How to collect and voucher mushroom specimens for science (Fungal Diversity Survey)

 Tips for mushroom photography (Central Texas Mycology)

How to Identify A Mushroom

1. Observe the mushroom's basic form (use Mushroom Morphological Types, page 17).
2. Study the specific morphological features of the specimen.
3. Understand the mushroom's habitat and environmental context.
4. Use your senses: sight, smell, hearing, touch, and taste (but don't swallow!).
5. Document the specimen by taking quality photos.
6. Upload pictures to an identification app, forum, or social media.
7. Compare your specimen(s) with books and online resources.
8. Collect, dissect, and examine the fruiting body.
9. Take a spore print and look at the spores under a microscope (optional).
10. Preserve the specimen by dehydrating for later analysis (optional).

How to Make a Spore Print

1. Remove the stem of the mushroom and place the cap gill/pore-side down on paper or aluminum foil.
2. Carefully remove cover and check the spore deposit.
3. Cover with a breathable container so air currents won't interfere with creating a clear spore deposit.
4. Check the color of the spore print and use for microscopy or cultivation, or just enjoy as art!

Mushroom Morphology Descriptive Terms

Cap shapes: conical, cylindrical, bell-shaped (campanulate), rounded (convex), egg-shaped (ovoid), flat (plane), seashell (conchate), pointed in the middle (umbonate), navel-shaped (umbilicate), funnel-shaped (infundibuliform).

Cap texture: smooth, cracked (aerolate), wet/slippery (viscid), concentrically ringed (zonate), warty wrinkled (rugose), fibrous (fibrillose), scaly tufted (floccose), peach fuzz (pubescent), woolly (tomentose), hairy (hispid), color change with moisture (hygrophanous), powdery white (pruinose), granular (granulose).

Cap edges (margins): striated (striate), wavy (undulating), fragments of veil (appendiculate), split, pleated/corrugated (plicate), hairy (tomentose).

Gill attachments: not attached (free), broadly attached (adnate), almost free (adnexed), notched at attachment (emarginate), running down stem (decurrent).

Gill edges: even, serrated, eroded, fringed (fimbriate), different colors on edge of gill typically matching the cap (marginate).

Gill spacing: crowded, close, subdistant, distant.

Spore-bearing surfaces (hymenium): gills, pores, tubes, teeth, ridges, wrinkles, gleba, cup, flask.

Stem textures: ornamented, webbed (reticulation), spotted (stippled), granular (granulose), fibrous (fibrillose), marked with v-shapes (chevrons), deeply grooved (lacunose), vertically striated (longitudinally striate).

Base: sac-like (saccate), bulbous, equal tapered, flaring, ridged (marginate).

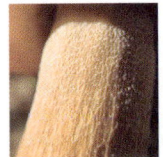

Reticulation
on stem

Decurrent
ridges

Squamulose
cap

Striate
cap edges

Mushroom Morphological Types

Note: The Latin root *oid* means "having the likeness of." In mycological convention, this refers to informal groupings of similar-looking mushrooms that are not necessarily related.

Gilled Cap-and-Stem Mushrooms (Agaricoids)

Chanterelle-like Mushrooms (Cantharelloids)

Oyster-like Mushrooms (Pleurotoids)

Tooth Mushrooms (Hydnoids)

Boletes (Boletoid)

Polypores with/without stem (Poroid)

Crusts & Parchments (Corticioids, Stereoids)

Corals & Clubs (Clavarioids)

Jelly Mushrooms (Tremelloids)

Puffballs, Stinkhorns, Earthstars, Bird's Nests (Gasteroids)

Truffles (Hypogeous Gasteroids)

Pouch Mushrooms (Secotioids)

Cups & Discs (Cupulates, Discoids)

Morels & Allies (Morchelloid)

Earthtongues & Allies (Glossioid)

Carbon & Flask Fungi (Perithecioid)

Figure 2. These morphological types represent most of the basic shapes and forms that mushrooms assume. In some cases, these groupings are genetically related, but not always. Compare the genera listed in these morphological categories with the taxonomic tree.

Dr. Fun Guy's Note on Mushroom Identification

Personally, I have never had much success using dichotomous keys, which are identification tools that rely on following binary choices for specific morphological traits, leading to a particular category of mushroom. Instead, I prefer the approach of comparing photos of mushrooms until I see something similar to my specimen. My strategy is to create visual fingerprints in my mind and use them to help recognize specific traits and mushroom forms. However, this can be hard to do without a reference, so here is a simplified pictorial key that represents the general logic I follow when assessing an unknown mushroom specimen. I hope this flowchart will be useful in helping you quickly recognize and categorize your finds.

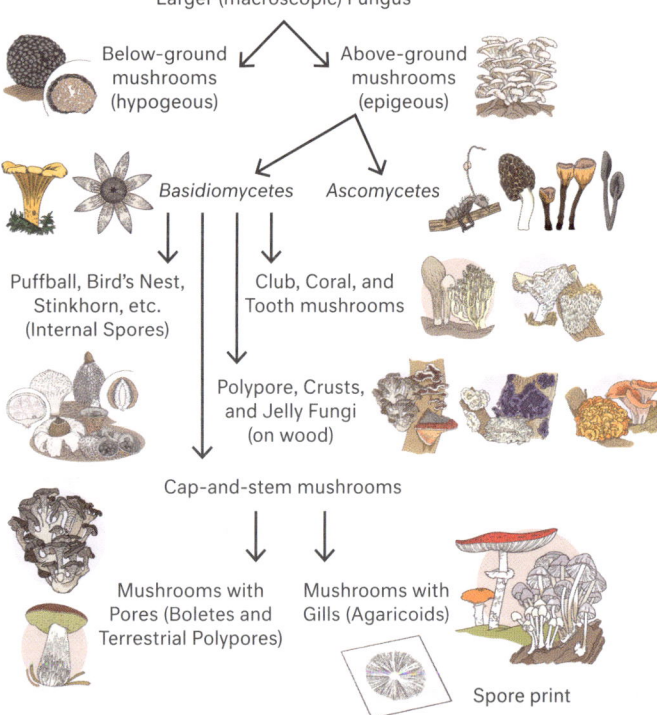

Figure 3. Visual flowchart for recognizing different forms of mushrooms.

Mushroom Edibility Desirability Ratings

**INCREDIBLE
(5 STAR)**

Truffle (*Tuber*)

**GREAT
(4 STAR)**

Chanterelle
(*Cantharellus*)

**GOOD
(3 STAR)**

Oyster Mushrooms
(*Pleurotus*)

**FAIR/MEDIOCRE
(2 STAR)**

Slippery Jack
(*Suillus*)

**NONTOXIC/
TECHNICALLY
EDIBLE (1 STAR)**

Bird's Nest Fungi
(*Crucibulum*)

**SPECIAL
PREPARATION/
MEDICINAL**

Turkey Tail
(*Trametes*)

For most foragers, Incredible, Great, and Good (5–3 Star) mushrooms are the primary focus. In general, mushrooms taste the way they smell. If something smells good, chances are it will taste good and vice versa.

Some intrepid foragers may also seek out mushrooms that are initially toxic but can be safely consumed after undergoing special preparation. Many medicinal mushrooms also require special preparation for consumption. **Toxic look-alikes are listed in the following sections.**

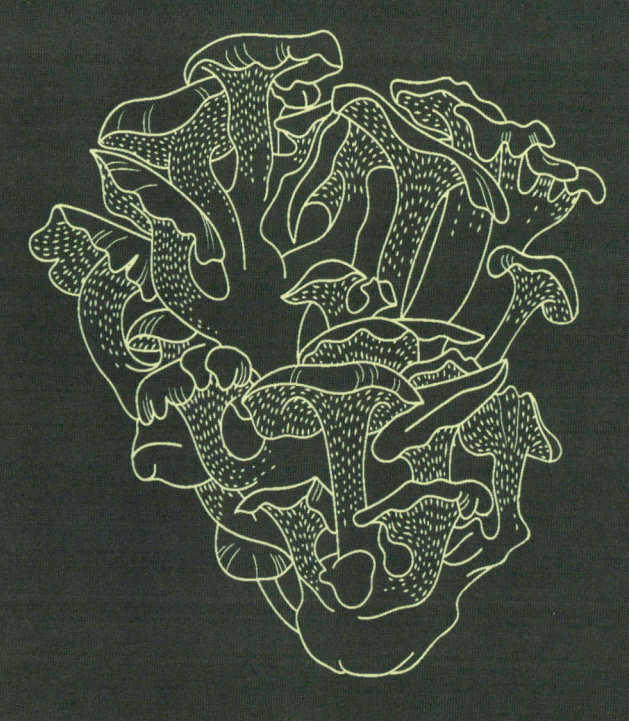

CHAPTER 1

Edible Polypores

Sulphur Shelf
(*Laetiporus sulphureus*)

White-Pored
Chicken of the Woods
(*Laetiporus cincinnatus*)

Western
Chicken of the Woods
(*Laetiporus gilbertsonii*)

More pictures
and info here:

Chicken of the Woods
(*Laetiporus*)

Taxonomy: *Basidiomycota, Agaricomycetes, Polyporales, Laetiporaceae*

Edibility/Toxicity: Great Edible (4 Stars)

Habitat (Association): Grows on wood (Broadleaf/Conifer) or at the base of oak trees

Trophic Mode(s): Parasitic, Saprotrophic (Brown Rot)

Spores: White (Basidiospore)

Appearance: Yellow-orange fleshy shelves or pinkish rosettes (fading to a pale beige as they mature)

Notable North American Species:
Broadleaf associated:
Laetiporus cincinnatus, gilbertsonii, sulphureus
Conifer associated:
L. coniferacola, huroniensis, montanus

Toxic Look-Alikes: Cinnamon bracket (*Hapalopilus rutilans*), jack-o'-lantern mushrooms (*Omphalotus*)

Safety Notes: Toxic when raw; always cook thoroughly. Roughly one in ten people will experience some degree of GI upset when consuming chicken of the woods. Be cautious when trying this mushroom for the first time and do not overindulge!

Date: _____

Location: _____

_____

_____

_____

Surrounding habitat/context: _____

_____

_____

_____

Morphology: _____

_____

_____

_____

Notes on color and texture: _____

_____

_____

_____

Smell: _____

_____

_____

_____

Other observations: _____

_____

_____

_____

_____

Maitake *(Grifola frondosa)*

Maitake *(Grifola frondosa)* at the base of an oak tree

Underside of maitake fronds

Maitake cluster

Alternate Names: Hen of the Woods, Sheep's / Ram's Head Polypore

Taxonomy: *Basidiomycota, Agaricomycetes, Polyporales, Grifolaceae*

Edibility: Incredible Edible (5 Stars)

Habitat (Association): Grows on ground (Broadleaf) at the base of mature oak trees, east of the Rocky Mountains

Trophic Mode(s): Parasitic, Saprotrophic (White Rot)

Spores: White (Basidiospore)

Appearance: Large gray to brown rosettes with leafy fronds sprouting outward from a central base root; small white angular pores on the underside

Look-Alikes: Black-staining polypore *(Meripilus)*

More pictures and info here:

Date: _____

Location: _____

_____

_____

_____

Surrounding habitat/context: _____

_____

_____

_____

Morphology: _____

_____

_____

_____

Notes on color and texture: _____

_____

_____

_____

Smell: _____

_____

_____

_____

Other observations: _____

_____

_____

_____

_____

Beefsteak (*Fistulina*) shelves growing at the base of a chinquapin oak

Kidney-shaped beefsteak in hand

Beefsteak cross section

Beefsteak Mushrooms
(*Fistulina*)

Alternate Name: Ox Tongue Fungus

Taxonomy: *Basidiomycota*, *Agaricomycetes*, *Agaricales*, *Fistulinaceae*

Edibility/Toxicity: Good Edible (3 Stars); edible when raw

Habitat (Association): Grows on wood (Broadleaf), out of the base of trees

Trophic Mode(s): Parasitic, Saprotrophic (Brown Rot)

Spores: Pinkish yellow to pinkish brown (Basidiospore)

Appearance: Large reddish-brown shelf-like ovals, topped when young by a layer of squishy red jelly; when mature, they resemble a kidney or liver

Toxic Look-Alikes: Immature cinnamon bracket (*Hapalopilus rutilans*)

More pictures and info here:

Date: _____

Location: _____

_____

_____

_____

Surrounding habitat/context: _____

_____

_____

_____

Morphology: _____

_____

_____

_____

Notes on color and texture: _____

_____

_____

_____

Smell: _____

_____

_____

_____

Other observations: _____

_____

_____

_____

_____

Cauliflower Mushrooms
(*Sparassis*)

Cauliflower Mushroom
(*Sparassis radicata*)

Eastern
Cauliflower Mushroom
(*Sparassis spathulata*)

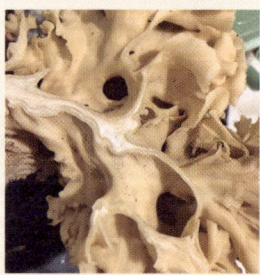

Cauliflower cross section

Taxonomy: *Basidiomycota, Agaricomycetes, Polyporales, Sparassidaceae*

Edibility/Toxicity: Incredible Edible (5 Stars)

Habitat (Association): Grows on ground, near the base of conifers like pine and Douglas fir on the West Coast; is most common in association with broadleaf hardwoods like oak in the rest of North America

Trophic Mode(s): Parasitic, Saprotrophic (Brown Rot)

Spores: Whitish (Basidiospore)

Appearance: Pale-colored, spatula-like fronds or leafy ruffled lobes, resembling a pile of egg noodles more than a mushroom

Notable North American Species: *Sparassis americana*, *radicata*, *spathulata*

More pictures and info here:

Date: _____

Location: _____

_____

_____

_____

Surrounding habitat/context: _____

_____

_____

_____

Morphology: _____

_____

_____

_____

Notes on color and texture: _____

_____

_____

_____

Smell: _____

_____

_____

_____

Other observations: _____

_____

_____

_____

_____

Lion's Mane
(*Hericium erinaceus*)

Coral Tooth
(*Hericium coralloides*)

Bear's Head Tooth
(*Hericium americanum*)

Lion's Mane & Relatives (*Hericium*)

Alternate Names: Coral Tooth, Bear's Head, Bearded Tooth

Taxonomy: *Basidiomycota, Agaricomycetes, Russulales, Hericiaceae*

Edibility/Toxicity: Incredible Edible (5 Stars)

Habitat (Association): Grows directly from the wood of live trees (most often hardwoods, but can also be found on fir trees)

Trophic Mode(s): Parasitic, Saprotrophic (White Rot)

Spores: White (Basidiospore)

Appearance: Lion's mane forms rounded balls, while other *Hericium* species tend to form more branched coral-like clusters or looser toothy chunks.

Notable North American Species: *Hericium abietis, cirrhatum, coralloides, erinaceus*

Look-Alikes: *Irpex rosettiformis, Daleomyces phillipsii*

More pictures and info here:

Date: _____

Location: _____

_____

_____

_____

Surrounding habitat/context: _____

_____

_____

_____

Morphology: _____

_____

_____

_____

Notes on color and texture: _____

_____

_____

_____

Smell: _____

_____

_____

_____

Other observations: _____

_____

_____

_____

_____

Black-Staining Polypores
(*Meripilus*)

Berkeley's Polypores
(*Bondarzewia*)

Sheep / Knight Polypores
(*Albatrellus*)

Pheasant Backs / Dryad's Saddles (*Cerioporus*)

Resinous Polypores (*Ischnoderma*)

Umbrella Polypore (*Polyporus umbellatus*)

Mushroom Common Name: _____

_____

Taxonomy: _____

_____

Edibility/Toxicity: _____

Habitat (Association): _____

_____

Trophic Mode(s): _____

_____

Spores: _____

Appearance: _____

_____

Notable North American Species: _____

_____

Toxic Look-Alikes: _____

Look-alikes: _____

Date: _____ Location: _____

Surrounding habitat/context: _____

_____

Morphology: _____

_____

Notes on color and texture: _____

_____

Smell: _____

_____

Other observations: _____

_____

Mushroom Common Name: _____

_____

Taxonomy: _____

_____

Edibility/Toxicity: _____

Habitat (Association): _____

_____

Trophic Mode(s): _____

_____

Spores: _____

Appearance: _____

_____

Notable North American Species: _____

_____

Toxic Look-Alikes: _____

Look-alikes: _____

Date: _____ Location: _____

Surrounding habitat/context: _____

_____

Morphology: _____

_____

Notes on color and texture: _____

_____

Smell: _____

_____

Other observations: _____

_____

Mushroom Common Name: _____

_____

Taxonomy: _____

_____

Edibility/Toxicity: _____

Habitat (Association): _____

_____

Trophic Mode(s): _____

_____

Spores: _____

Appearance: _____

_____

Notable North American Species: _____

_____

Toxic Look-Alikes: _____

Look-alikes: _____

Date: _____ Location: _____

Surrounding habitat/context: _____

_____

Morphology: _____

_____

Notes on color and texture: _____

_____

Smell: _____

_____

Other observations: _____

_____

Mushroom Common Name: _____

_____

Taxonomy: _____

_____

Edibility/Toxicity: _____

Habitat (Association): _____

_____

Trophic Mode(s): _____

_____

Spores: _____

Appearance: _____

_____

Notable North American Species: _____

_____

Toxic Look-Alikes: _____

Look-alikes: _____

Date: _____ Location: _____

Surrounding habitat/context: _____

_____

Morphology: _____

_____

Notes on color and texture: _____

_____

Smell: _____

_____

Other observations: _____

_____

Mushroom Common Name: _____

_____

Taxonomy: _____

_____

Edibility/Toxicity: _____

Habitat (Association): _____

_____

Trophic Mode(s): _____

_____

Spores: _____

Appearance: _____

_____

Notable North American Species: _____

_____

Toxic Look-Alikes: _____

Look-alikes: _____

Date: _____ Location: _____

Surrounding habitat/context: _____

_____

Morphology: _____

_____

Notes on color and texture: _____

_____

Smell: _____

_____

Other observations: _____

_____

Mushroom Common Name: _____

_____

Taxonomy: _____

_____

Edibility/Toxicity: _____

Habitat (Association): _____

_____

Trophic Mode(s): _____

_____

Spores: _____

Appearance: _____

_____

Notable North American Species: _____

_____

Toxic Look-Alikes: _____

Look-alikes: _____

Date: _____ Location: _____

Surrounding habitat/context: _____

_____

Morphology: _____

_____

Notes on color and texture: _____

_____

Smell: _____

_____

Other observations: _____

_____

CHAPTER 2

Edible Boletes

California King Bolete
(*Boletus edulis* var. *grandedulis*)

Ruby Porcini
(*Boletus rubriceps*)

Tan King Bolete
(*Boletus variipes*)

King Boletes
(*Boletus edulis* group)

Alternate Names: Porcini, Ceps, Penny Bun

Taxonomy: *Basidiomycota*, *Agaricomycetes*, *Boletales*, *Boletaceae*

Edibility/Toxicity: Great Edible (4 Stars); edible when raw

Habitat (Association): Grows on ground (Broadleaf/Conifer)

Trophic Mode(s): Ectomycorrhizal

Spores: Olive brown (Basidiospore)

Appearance: Large and fleshy with a thick stem and a domed cap that can be gray, tan, copper, reddish, or dark brown

Notable North American Species: *Boletus barrowsii*, *chippewaensis*, *edulis*, *regineus*, *rex-veris*, *rubriceps*, *variipes*

Toxic Look-Alikes: False king bolete (*Boletus huronensis*)

More pictures and info here:

Date: _____

Location: _____

_____

_____

_____

Surrounding habitat/context: _____

_____

_____

_____

Morphology: _____

_____

_____

_____

Notes on color and texture: _____

_____

_____

_____

Smell: _____

_____

_____

_____

Other observations: _____

_____

_____

_____

_____

_____

Pink-Capped
Oak Butter Bolete
(*Butyriboletus querciregius*)

Butter Bolete
(*Butyriboletus primiregius*)

Brown Butter Bolete
(*Butyriboletus persolidus*)

Butter Boletes
(*Butyriboletus*)

Taxonomy: *Basidiomycota, Agaricomycetes, Boletales, Boletaceae*

Edibility/Toxicity: Incredible Edible (5 Stars); edible when raw

Habitat (Association): Grows on ground (Broadleaf/Conifer), often buried under leaf litter

Trophic Mode(s): Ectomycorrhizal

Spores: Olive brown (Basidiospore)

Appearance: A big, smooth, reddish-pink to light brown cap; a thick yellowish stem with well-defined reticulation; flesh that will slowly and erratically stain blue when cut

Notable North American Species: *Butyriboletus abieticola, appendiculatus, autumniregius, brunneus, persolidus, primiregius, pulchriceps, querciregius, roseopurpureus*

Toxic Look-Alikes: Satan's boletes (*Rubroboletus*)

Look-Alikes: Bitter boletes (*Caloboletus*), red cracking boletes (*Xerocomellus*)

More pictures
and info here:

Date: _____

Location: _____

_____

_____

_____

Surrounding habitat/context: _____

_____

_____

_____

Morphology: _____

_____

_____

_____

Notes on color and texture: _____

_____

_____

_____

Smell: _____

_____

_____

_____

Other observations: _____

_____

_____

_____

_____

_____

Short-Stalked Jack
(*Suillus brevipes*)

Suillus pores and
partial veil

Suillus pungens
with *Chroogomphus*

More pictures
and info here:

Jacks (*Suillus*)

Alternate Names: Slippery Jack, Chicken Fat Mushroom

Taxonomy: *Basidiomycota*, *Agaricomycetes*, *Boletales*, *Suillaceae*

Edibility/Toxicity: Mediocre to Great Edible (2-4 Stars)

Habitat (Association): Grows on ground (Conifer)

Trophic Mode(s): Ectomycorrhizal

Spores: Olive brown (Basidiospore)

Appearance: Smooth and slimy caps and yellow pores, in a wide variety of sizes, shapes, and colors

Notable North American Species: *Suillus americanus*, *brevipes*, *caerulescens*, *granulatus*, *lakei*, *luteus*, *ponderosus*, *pungens*, *salmonicolor*, *spraguei*, *tomentosus*, *weaverae*

Look-Alikes: Golden pore boletes (*Aureoboletus*), powdery sulfur boletes (*Pulveroboletus*), wood boletes (*Buchwaldoboletus*)

Safety Note: The slippery nature of jacks can sometimes cause a rash for sensitive individuals; consider wearing gloves when handling. Peel off the slime membrane on the cap when cooking to mitigate the chance of GI upset.

Date: _____

Location: _____

_____

_____

_____

Surrounding habitat/context: _____

_____

_____

_____

Morphology: _____

_____

_____

_____

Notes on color and texture: _____

_____

_____

_____

Smell: _____

_____

_____

_____

Other observations: _____

_____

_____

_____

_____

_____

Scaber Stalks (*Leccinum*)

Manzanita Scaber Stalk
(*Leccinum manzanitae*)

Brown Scaber Stalk
(*Leccinum scabrum*)

Aspen Scaber Stalk
(*Leccinum insigne*)

Alternate Names: Birch or Aspen Bolete

Taxonomy: *Basidiomycota, Agaricomycetes, Boletales, Boletaceae*

Edibility/Toxicity: Good Edible (3 Stars)

Habitat (Association): Grows on ground (Broadleaf/Conifer) alongside a specific tree type (the best-known species grow with birch trees and are commonly known as birch boletes)

Trophic Mode(s): Ectomycorrhizal

Spores: Olive brown (Basidiospore)

Appearance: A compact, rounded cap—light yellow, orangey-red, or dark brown—with a wrinkled or suede-like texture; uniformly thick stem covered in black or brown protruding glandular dots

Notable North American Species: *Leccinum albostipitatum, aurantiacum, duriusculum, holopus, insigne, longicurvipes, manzanitae, scabrum*

Safety Notes: Avoid orange-capped species commonly found in the Rocky Mountains in association with aspen trees, as there have been reports of poisonings associated with these (specifically *L. insigne*). Always cook *Leccinum* thoroughly to avoid any chance of GI upset.

More pictures and info here:

Date: _____

Location: _____

_____

_____

_____

Surrounding habitat/context: _____

_____

_____

_____

Morphology: _____

_____

_____

_____

Notes on color and texture: _____

_____

_____

_____

Smell: _____

_____

_____

_____

Other observations: _____

_____

_____

_____

_____

Golden Pore Boletes
(*Aureoboletus*)

Bicolor Bolete
(*Baorangia bicolor*)

Cornflower Bolete
(*Gyroporus cyanescens*)

Suede Boletes
(*Xerocomus*)

Red Cracking Boletes
(*Xerocomellus*)

Velvet Boletes
(*Tylopilus*)

Old Man of the Woods
(*Strobilomyces*)

Gilled Boletes
(*Phylloporus*)

Mushroom Common Name: _____

_____

Taxonomy: _____

_____

Edibility/Toxicity: _____

Habitat (Association): _____

_____

Trophic Mode(s): _____

_____

Spores: _____

Appearance: _____

_____

Notable North American Species: _____

_____

Toxic Look-Alikes: _____

Look-alikes: _____

Date: _____ Location: _____

Surrounding habitat/context: _____

_____

Morphology: _____

_____

Notes on color and texture: _____

_____

Smell: _____

_____

Other observations: _____

_____

Mushroom Common Name: _____

_____

Taxonomy: _____

_____

Edibility/Toxicity: _____

Habitat (Association): _____

_____

Trophic Mode(s): _____

_____

Spores: _____

Appearance: _____

_____

Notable North American Species: _____

_____

Toxic Look-Alikes: _____

Look-alikes: _____

Date: _____ Location: _____

Surrounding habitat/context: _____

_____

Morphology: _____

_____

Notes on color and texture: _____

_____

Smell: _____

_____

Other observations: _____

_____

Mushroom Common Name: _____

_____

Taxonomy: _____

_____

Edibility/Toxicity: _____

Habitat (Association): _____

_____

Trophic Mode(s): _____

_____

Spores: _____

Appearance: _____

_____

Notable North American Species: _____

_____

Toxic Look-Alikes: _____

Look-alikes: _____

Date: _____ Location: _____

Surrounding habitat/context: _____

_____

Morphology: _____

_____

Notes on color and texture: _____

_____

Smell: _____

_____

Other observations: _____

_____

Mushroom Common Name: _____

_____

Taxonomy: _____

_____

Edibility/Toxicity: _____

Habitat (Association): _____

_____

Trophic Mode(s): _____

_____

Spores: _____

Appearance: _____

_____

Notable North American Species: _____

_____

Toxic Look-Alikes: _____

Look-alikes: _____

Date: _____ Location: _____

Surrounding habitat/context: _____

_____

Morphology: _____

_____

Notes on color and texture: _____

_____

Smell: _____

_____

Other observations: _____

_____

Mushroom Common Name: _____

_____

Taxonomy: _____

_____

Edibility/Toxicity: _____

Habitat (Association): _____

_____

Trophic Mode(s): _____

_____

Spores: _____

Appearance: _____

_____

Notable North American Species: _____

_____

Toxic Look-Alikes: _____

Look-alikes: _____

Date: _____ Location: _____

Surrounding habitat/context: _____

_____

Morphology: _____

_____

Notes on color and texture: _____

_____

Smell: _____

_____

Other observations: _____

_____

Mushroom Common Name: _____

_____

Taxonomy: _____

_____

Edibility/Toxicity: _____

Habitat (Association): _____

_____

Trophic Mode(s): _____

_____

Spores: _____

Appearance: _____

_____

Notable North American Species: _____

_____

Toxic Look-Alikes: _____

Look-alikes: _____

Date: _____ Location: _____

Surrounding habitat/context: _____

_____

Morphology: _____

_____

Notes on color and texture: _____

_____

Smell: _____

_____

Other observations: _____

_____

Mushroom Common Name: _____

_____

Taxonomy: _____

_____

Edibility/Toxicity: _____

Habitat (Association): _____

_____

Trophic Mode(s): _____

_____

Spores: _____

Appearance: _____

_____

Notable North American Species: _____

_____

Toxic Look-Alikes: _____

Look-alikes: _____

Date: _____ Location: _____

Surrounding habitat/context: _____

_____

Morphology: _____

_____

Notes on color and texture: _____

_____

Smell: _____

_____

Other observations: _____

_____

Mushroom Common Name: _____

_____

Taxonomy: _____

_____

Edibility/Toxicity: _____

Habitat (Association): _____

_____

Trophic Mode(s): _____

_____

Spores: _____

Appearance: _____

_____

Notable North American Species: _____

_____

Toxic Look-Alikes: _____

Look-alikes: _____

Date: _____ Location: _____

Surrounding habitat/context: _____

_____

Morphology: _____

_____

Notes on color and texture: _____

_____

Smell: _____

_____

Other observations: _____

_____

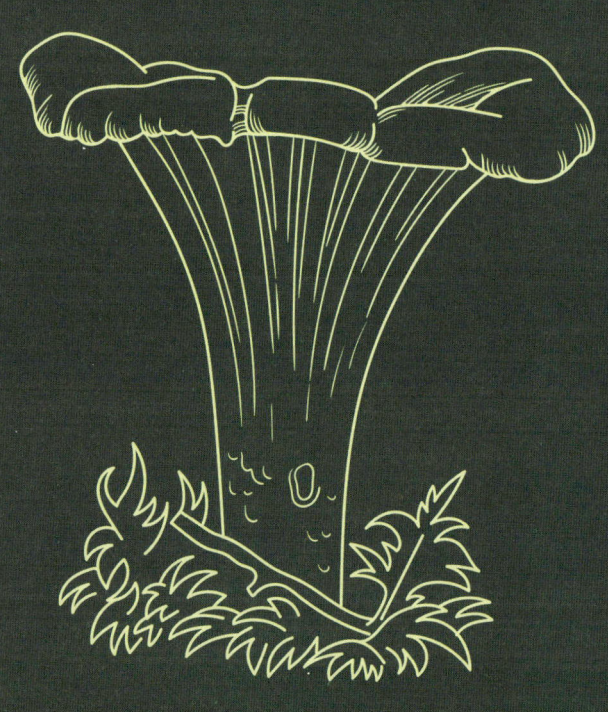

CHAPTER 3

Easy Edibles

Cascade Chanterelle
(*Cantharellus cascadensis*)

White Chanterelle
(*Cantharellus subalbidus*)

"False" chanterelle (left) vs.
Golden Pacific Chanterelle
(*Cantharellus formosus*)

More pictures
and info here:

"True" Chanterelles
(*Cantharellus*)

Taxonomy: *Basidiomycota, Agaricomycetes, Cantharellales, Cantharellaceae*

Edibility/Toxicity: Great Edible (4 Stars)

Habitat (Association): Grows directly from ground (Broadleaf/Conifer) with a wide range of plant hosts and habitats

Trophic Mode(s): Ectomycorrhizal

Spores: White (Basidiospore)

Appearance: Funnel-shaped in eye-catching colors (including creamy white, bright yellow, golden orange, red, and iridescent pink) with distinctive blunt decurrent ridges running down the stem

Notable North American Species: *Cantharellus appalachiensis, californicus, cascadensis, cinnabarinus, enelensis, flavus, flavolateritius, formosus, lateritius, minor, roseocanus, subalbidus*

Toxic Look-Alikes: Jack-o'-lantern mushrooms (*Omphalotus*), scaly "chanterelle" (*Turbinellus floccosus*)

Look-Alikes: Blue "chanterelles" (*Polyozellus*), false "chanterelle" (*Hygrophoropsis aurantiaca*), pig's ear (*Gomphus clavatus*), spikes (*Chroogomphus*), waxcaps (*Hygrocybe*)

Date: _____

Location: _____

_____

_____

_____

Surrounding habitat/context: _____

_____

_____

_____

Morphology: _____

_____

_____

_____

Notes on color and texture: _____

_____

_____

_____

Smell: _____

_____

_____

_____

Other observations: _____

_____

_____

_____

_____

_____

Yellowfeet and Black Trumpets (*Craterellus*)

Winter Chanterelle or Yellowfoot (*Craterellus tubaeformis* var. *pacifica*)

Flame Trumpet (*Craterellus ignicolor*)

Black Trumpets (*Craterellus calicornucopiodes*)

Alternate Names: Funnel Chanterelle, Horn of Plenty, Trumpet of the Dead

Taxonomy: *Basidiomycota, Agaricomycetes, Cantharellales, Cantharellaceae*

Edibility/Toxicity: Great Edible (4 Stars)

Habitat (Association): Grows on ground (Broadleaf/Conifer) in mossy areas, trailside berms, or thick layers of duff

Trophic Mode(s): Ectomycorrhizal

Spores: White to yellowish (Basidiospore)

Appearance: Small, with a hollow central stem and wrinkly decurrent ridges (false gills). Yellowfeet have a wider flaring cap, while black trumpets have a more vase-like shape.

Notable North American Species: *Craterellus atrocinereus, calicornucopioides, cornucopioides, fallax, ignicolor, lutescens, odoratus, tubaeformis*

Look-alikes: Earthfans (*Thelephora*), waxy caps (*Hygrocybe*)

More pictures and info here:

Date: _____

Location: _____

_____

_____

_____

Surrounding habitat/context: _____

_____

_____

_____

Morphology: _____

_____

_____

_____

Notes on color and texture: _____

_____

_____

_____

Smell: _____

_____

_____

_____

Other observations: _____

_____

_____

_____

_____

_____

Bellybutton Hedgehog
(*Hydnum oregonense*)

Wood Hedgehog (*Hydnum washingtonianum*)

Handful of hedgehog mushrooms

Hedgehog Mushrooms
(*Hydnum*)

Taxonomy: *Basidiomycota, Agaricomycetes, Cantharellales, Hydnaceae*

Edibility/Toxicity: Incredible Edible (5 Stars)

Habitat (Association): Grows on ground (Broadleaf/Conifer)

Trophic Mode(s): Ectomycorrhizal

Spores: White (Basidiospore)

Appearance: A well-defined stem and white, pink, gold, or brown cap with distinctive small teeth or spines on its underside

Notable North American Species: *Hydnum albidum, aerostatisporum, oregonense, repandum, rufescens, subolympicum, washingtonianum, umbilicatum*

Look-Alikes: (When seen from below): shingled hedgehogs (*Sarcodon*), bitter tooth (*Hydnellum*). (When seen from above): chanterelles, meadow waxy caps (*Cuphophyllus*).

More pictures and info here:

Date: _____

Location: _____

_____

_____

_____

Surrounding habitat/context: _____

_____

_____

_____

Morphology: _____

_____

_____

_____

Notes on color and texture: _____

_____

_____

_____

Smell: _____

_____

_____

_____

Other observations: _____

_____

_____

_____

_____

_____

Meadow Puffball
(*Lycoperdon pratense*)

Lycoperdon spines

Sculpted Puffball
(*Calvatia sculpta*)

Puffballs (*Lycoperdaceae*)

Taxonomy: *Basidiomycota, Agaricomycetes, Agaricales, Lycoperdaceae*

Edibility/Toxicity: Good Edible (3 Stars)

Habitat (Association): Grows on ground (Soil/Grass), although light brown puffballs can grow on decaying wood

Trophic Mode(s): Saprotrophic (Composter)

Spores: Brown (Basidiospore)

Appearance: White, gray, or light brown; with stalks or entirely round; surface may be completely smooth, spiky, wrinkled, or like piped meringue or soccer-ball stitching

Notable North American Genera: *Apioperdon, Bovista, Calvatia, Handkea, Mycenastrum, Lycoperdon*

Look-Alikes: Immature *Amanita* eggs (*Amanita ovoidea*), earthballs (*Scleroderma*), dyeballs (*Pisolithus*)

Safety Notes: Inhaling a large amount of the mature spores can lead to respiratory infections. Avoid gathering in urban and industrial areas, as puffballs can bioaccumulate heavy-metal contaminants like mercury.

More pictures and info here:

Date: _____

Location: _____

_____

_____

_____

Surrounding habitat/context: _____

_____

_____

_____

Morphology: _____

_____

_____

_____

Notes on color and texture: _____

_____

_____

_____

Smell: _____

_____

_____

_____

Other observations: _____

_____

_____

_____

_____

_____

Lobster Mushroom
(*Hypomyces lactifluorum*)

Lobster mushroom
(*Hypomyces lactifluorum*)
in hand

Lobster mushroom
color gradient

Yellow-green Russula Mold
(*Hypomyces luteovirens*)
is also a good edible.

Taxonomy: (Base Mushroom) *Basidiomycota, Agaricomycetes, Russulales, Russulaceae*

Taxonomy: (Parasite) *Ascomycota, Sordariomycetes, Hypocreales, Hypocreaceae*

Edibility/Toxicity: Good Edible (3 Stars)

Habitat (Association): Grows on ground (Broadleaf/Conifer)

Trophic Mode(s): Ectomycorrhizal /Parasitic

Spores: White (Ascospore)

Appearance: Bright reddish-orange distorted shapes created when this edible mold takes over another mushroom species; no gills visible under the cap, the area being completely covered and smoothed over

Safety Note: While *Hypomyces* can colonize inedible *Russula/ Lactarius*, there have been no reports of poisonings due to consumption of this mushroom in the last hundred years.

More pictures
and info here:

Date: _____

Location: _____

_____

_____

_____

Surrounding habitat/context: _____

_____

_____

_____

Morphology: _____

_____

_____

_____

Notes on color and texture: _____

_____

_____

_____

Smell: _____

_____

_____

_____

Other observations: _____

_____

_____

_____

_____

_____

Honorable Mentions

Pig's Ear
(*Gomphus clavatus*)

Hawk's Wing
(*Sarcodon imbricatus*)

Splitgill
(*Schizophyllum commune*)

Wood Ears
(*Auricularia*)

Mushroom Common Name: _____

Taxonomy: _____

Edibility/Toxicity: _____

Habitat (Association): _____

Trophic Mode(s): _____

Spores: _____

Appearance: _____

Notable North American Species: _____

Toxic Look-Alikes: _____

Look-alikes: _____

Date: _____ Location: _____

Surrounding habitat/context: _____

Morphology: _____

Notes on color and texture: _____

Smell: _____

Other observations: _____

Mushroom Common Name: _____

_____

Taxonomy: _____

_____

Edibility/Toxicity: _____

Habitat (Association): _____

_____

Trophic Mode(s): _____

_____

Spores: _____

Appearance: _____

_____

Notable North American Species: _____

_____

Toxic Look-Alikes: _____

Look-alikes: _____

Date: _____ Location: _____

Surrounding habitat/context: _____

_____

Morphology: _____

_____

Notes on color and texture: _____

_____

Smell: _____

_____

Other observations: _____

_____

Mushroom Common Name: _____

_____

Taxonomy: _____

_____

Edibility/Toxicity: _____

Habitat (Association): _____

_____

Trophic Mode(s): _____

_____

Spores: _____

Appearance: _____

_____

Notable North American Species: _____

_____

Toxic Look-Alikes: _____

Look-alikes: _____

Date: _____ Location: _____

Surrounding habitat/context: _____

_____

Morphology: _____

_____

Notes on color and texture: _____

_____

Smell: _____

_____

Other observations: _____

_____

Mushroom Common Name: _____

_____

Taxonomy: _____

_____

Edibility/Toxicity: _____

Habitat (Association): _____

_____

Trophic Mode(s): _____

_____

Spores: _____

Appearance: _____

_____

Notable North American Species: _____

_____

Toxic Look-Alikes: _____

Look-alikes: _____

Date: _____ Location: _____

Surrounding habitat/context: _____

_____

Morphology: _____

_____

Notes on color and texture: _____

_____

Smell: _____

_____

Other observations: _____

_____

CHAPTER 4

Intermediate Edibles

Shaggy Mane Inkcaps
(*Coprinus comatus*)

Big pile of
prime shaggy manes

Melted (deliquesced)
shaggy manes

Shaggy Mane Inkcap
(*Coprinus comatus*)

Alternate Name: Lawyer's Wig

Taxonomy: *Basidiomycota, Agaricomycetes, Agaricales, Agaricaceae*

Edibility/Toxicity: Great Edible (4 Stars)

Habitat (Association): Grows on ground (Grass/Gravel) in well-composted organic material and disturbed areas, often along roadsides and paths or in woodchips

Trophic Mode(s): Saprotrophic (Composter)

Spores: Black (Basidiospore)

Appearance: White, egg-shaped ovals with fast-growing, thin, hollow stems and elongated tubular caps adorned in shaggy brownish scales

Look-Alikes: Inkcaps (*Coprinopsis*), parasols (*Chlorophyllum/ Macrolepiota*), stalked puffballs (*Podaxis*)

Safety Note: Unlike other inkcaps, shaggy manes do not contain coprine. Thus, they are safe to consume with alcohol. However, other inkcaps can look relatively similar so be especially certain of your identification if cooking and consuming these with alcohol.

More pictures
and info here:

Date: _____

Location: _____

_____

_____

_____

Surrounding habitat/context: _____

_____

_____

_____

Morphology: _____

_____

_____

_____

Notes on color and texture: _____

_____

_____

_____

Smell: _____

_____

_____

_____

Other observations: _____

_____

_____

_____

_____

_____

Honey Mushrooms
(*Armillaria mellea*)

Honey Mushrooms
(*Armillaria gallica*)

Ringless Honey
Mushrooms
(*Desarmillaria caespitosa*)

Honey Mushrooms
(*Armillaria*)

Taxonomy: *Basidiomycota, Agaricomycetes, Agaricales, Physalacriaceae*

Edibility/Toxicity: Good Edible (3 Stars)

Habitat (Association): Grows on ground (Broadleaf/Conifer) around the base of dead or dying trees

Trophic Mode(s): Parasitic, Saprotrophic (White Rot)

Spores: White (Basidiospore)

Appearance: Bright yellow or brown cap, sometimes with a distinctive ring around the stalk

Notable North American Species: *Armillaria borealis, gallica, mellea, nabsnona, ostoyae, sinapina, tabescens*

Toxic Look-Alikes: Jack-o'-lantern mushrooms (*Omphalotus*), scalycaps (*Pholiota*), rustgills (*Gymnopilus*)

Safety Note: Honey mushrooms are quite slimy when cooked—to avoid potential GI upset, boil them for five to ten minutes before further cooking to mitigate the sliminess.

More pictures
and info here:

Date: _____

Location: _____

_____

_____

_____

Surrounding habitat/context: _____

_____

_____

_____

Morphology: _____

_____

_____

_____

Notes on color and texture: _____

_____

_____

_____

Smell: _____

_____

_____

_____

Other observations: _____

_____

_____

_____

_____

Cluster of oyster mushrooms (*Pleurotus*)

View of oyster mushroom gills

Oyster mushrooms on log

Oyster Mushrooms
(*Pleurotus*)

Taxonomy: *Basidiomycota, Agaricomycetes, Agaricales, Pleurotaceae*

Edibility/Toxicity: Good Edible (3 Stars)

Habitat (Association): Grows on wood (Broadleaf/Conifer), usually from the side of trees or fallen logs

Trophic Mode(s): Saprotrophic (White Rot)

Spores: White (Basidiospore)

Appearance: Various colors on top; well-defined creamy white gills on the underside, running down the thick offset stem

Notable North American Species: *Pleurotus citrinopileatus, djamor, dryinus, levis, ostreatus, populinus, pulmonarius*

Toxic Look-Alikes: Angel wing (*Pleurocybella porrigens*)

Look-Alikes: Soft slippers (*Crepidotus*), smoked oysterlings (*Resupinatus*), fall oysters (*Sarcomyxa*)

Safety Notes: Oyster mushrooms are mildly toxic when raw. All species should be cooked before consumption.

More pictures and info here:

Date: _____

Location: _____

_____

_____

_____

Surrounding habitat/context: _____

_____

_____

_____

Morphology: _____

_____

_____

_____

Notes on color and texture: _____

_____

_____

_____

Smell: _____

_____

_____

_____

Other observations: _____

_____

_____

_____

_____

_____

Blewit (*Clitocybe nuda*)
top and bottom

Blewit gills up close

Blewits on a branch

Blewits (*Clitocybe / Collybia nuda* group)

Taxonomy: *Basidiomycota, Agaricomycetes, Agaricales, Clitocybaceae*

Edibility/Toxicity: Good Edible (3 Stars)

Habitat (Association): Grows on ground (Soil/Duff), particularly in thick, moist organic matter such as leaves, straw, and compost; common along grassy embankments, in muddy drainage ditches, and beside paths and roadsides

Trophic Mode(s): Saprotrophic (Composter)

Spores: Light pink to lilac (Basidiospore)

Appearance: Intensely lilac purple (fading to light brown as they age), with a smooth stem and suede-like cap

Look-Alikes: Purple deceivers (*Laccaria*), purple webcaps (*Cortinarius*)

Safety Notes: Blewits are notorious for bioaccumulating heavy metals from the environment, lead especially. Avoid collecting near roads and areas with heavy traffic. Mildly toxic when raw, cook thoroughly.

More pictures and info here:

Date: _____

Location: _____

_____

_____

_____

Surrounding habitat/context: _____

_____

_____

_____

Morphology: _____

_____

_____

_____

Notes on color and texture: _____

_____

_____

_____

Smell: _____

_____

_____

_____

Other observations: _____

_____

_____

_____

_____

_____

Horse Mushroom
(*Agaricus arvensis*)

The Prince
(*Agaricus augustus*)

Mature Prince

Almond-Scented Agarics (*Agaricus*)

Alternate Names: Royal, Anise-Scented Agaric

Taxonomy: *Basidiomycota, Agaricomycetes, Agaricales, Agaricaceae*

Edibility/Toxicity: Great Edible (4 Stars)

Habitat (Association): Grows on ground (Duff/Grass); one species, the prince (*Agaricus augustus*), is common on the California coast among redwood duff

Trophic Mode(s): Saprotrophic (Composter)

Spores: Brown (Basidiospore)

Appearance: Large and robust; pinkish gills; cap and stem covered with woolly tufts (floccose), with yellowish ring around stem

Notable North American Species: *Agaricus arvensis, augustus, nanoaugustus, smithianus*

Toxic Look-Alikes: Yellow staining agarics (*A.* section *xanthodermatei*)

Look-Alikes: Parasols (*Chlorophyllum/Macrolepiota*)

Safety Notes: Avoid collecting along roads or in urban and industrial areas as agarics can bioaccumulate heavy metals.

More pictures and info here:

Date: _____

Location: _____

_____

_____

_____

Surrounding habitat/context: _____

_____

_____

_____

Morphology: _____

_____

_____

_____

Notes on color and texture: _____

_____

_____

_____

Smell: _____

_____

_____

_____

Other observations: _____

_____

_____

_____

_____

_____

Honorable Mentions

Sawgills / Trainwreckers
(*Neolentinus*)

Late Fall Oyster
(*Sarcomyxa serotina*)

Club Mushrooms
(*Clavariadelphus*)

Shrimp of the Woods
(*Entoloma abortivum*)

Mushroom Common Name: _____

_____

Taxonomy: _____

_____

Edibility/Toxicity: _____

Habitat (Association): _____

_____

Trophic Mode(s): _____

_____

Spores: _____

Appearance: _____

_____

Notable North American Species: _____

_____

Toxic Look-Alikes: _____

Look-alikes: _____

Date: _____ Location: _____

Surrounding habitat/context: _____

_____

Morphology: _____

_____

Notes on color and texture: _____

_____

Smell: _____

_____

Other observations: _____

_____

Mushroom Common Name: _____

_____

Taxonomy: _____

_____

Edibility/Toxicity: _____

Habitat (Association): _____

_____

Trophic Mode(s): _____

_____

Spores: _____

Appearance: _____

_____

Notable North American Species: _____

_____

Toxic Look-Alikes: _____

Look-alikes: _____

Date: _____ Location: _____

Surrounding habitat/context: _____

_____

Morphology: _____

_____

Notes on color and texture: _____

_____

Smell: _____

_____

Other observations: _____

_____

Mushroom Common Name: _____

_____

Taxonomy: _____

_____

Edibility/Toxicity: _____

Habitat (Association): _____

_____

Trophic Mode(s): _____

_____

Spores: _____

Appearance: _____

_____

Notable North American Species: _____

_____

Toxic Look-Alikes: _____

Look-alikes: _____

Date: _____ Location: _____

Surrounding habitat/context: _____

_____

Morphology: _____

_____

Notes on color and texture: _____

_____

Smell: _____

_____

Other observations: _____

_____

Mushroom Common Name: _____

_____

Taxonomy: _____

_____

Edibility/Toxicity: _____

Habitat (Association): _____

_____

Trophic Mode(s): _____

_____

Spores: _____

Appearance: _____

_____

Notable North American Species: _____

_____

Toxic Look-Alikes: _____

Look-alikes: _____

Date: _____ Location: _____

Surrounding habitat/context: _____

_____

Morphology: _____

_____

Notes on color and texture: _____

_____

Smell: _____

_____

Other observations: _____

_____

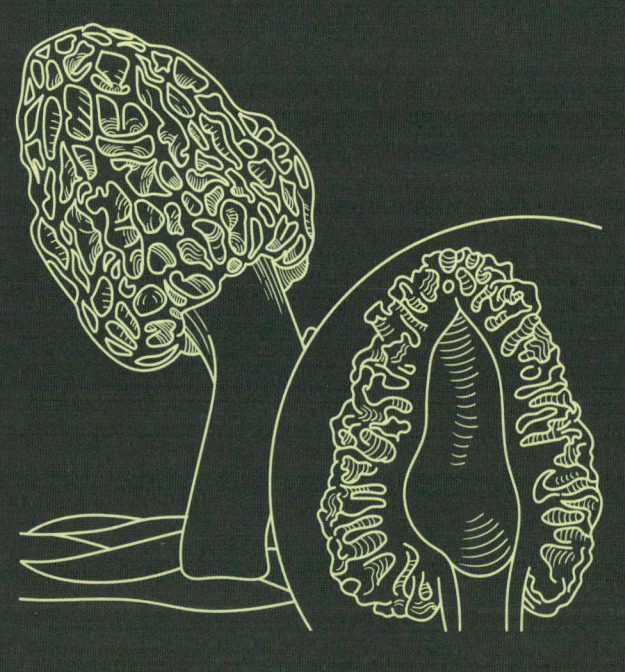

CHAPTER 5

Advanced Edibles

Yellow Morel
(*Morchella americana*)

Black Morels
(*Morchella sextelata*)

Woodchip Morel
(*Morchella rufobrunnea*)

"True" Morels (*Morchella*)

Taxonomy: *Ascomycota, Pezizomycetes, Pezizales, Morchellaceae*

Edibility/Toxicity: Incredible Edible (5 Stars)

Habitat (Association): Grows on ground (Broadleaf/Conifer); blonde morels in association with hardwoods on the East Coast and in the Midwest; black and gray morels in association with conifers after wildfires in the alpine West Coast; blushing morels in mulch and landscaping in cosmopolitan areas

Trophic Mode(s): Endophytic/ Saprotrophic (White Rot)

Spores: White (Ascospore)

Appearance: Conical, deeply pitted, honeycomb-like caps; hollow on the inside, with no internal folds or pith

Notable North American Species: *Morchella americana, angusticeps, brunnea, diminutiva, importuna, punctipes, rufobrunnea, snyderi, tomentosa, tridentina*

Look-Alikes: Lorchels (*Gyromitra*), elfin saddles (*Helvella*), thimble morel (*Verpa conica*)

Safety Notes: Raw morels are toxic and consumption can be fatal. Thorough cooking (ten to fifteen minutes) is required.

More pictures and info here:

Date: _____

Location: _____

_____

_____

_____

Surrounding habitat/context: _____

_____

_____

_____

Morphology: _____

_____

_____

_____

Notes on color and texture: _____

_____

_____

_____

Smell: _____

_____

_____

_____

Other observations: _____

_____

_____

_____

_____

_____

"True" Truffles (*Tuber*)

Big box of imported black truffles (*Tuber*)

Cross section of a black truffle

Truffles stored with rice to infuse aroma

Taxonomy: *Ascomycota, Pezizomycetes, Pezizales, Morchellaceae*

Edibility/Toxicity: Incredible Edible (5 Stars); edible when raw

Habitat (Association): Grows underground (Broadleaf/Conifer)

Trophic Mode(s): Ectomycorrhizal

Spores: Brown to blackish (Ascospore)

Appearance: Bumpy balls with a rough outer layer that looks cracked or warty (black, brown, or dingy beige); a distinctively ruffled internal structure. There are many types of truffles besides *Tuber* species. A few are quite tasty while many others are downright unpalatable. Let your nose guide you (appealing vs. unappealing aroma) and be cautious if you sample an unknown specimen.

More pictures and info here:

Date: _____

Location: _____

_____

_____

_____

Surrounding habitat/context: _____

_____

_____

_____

Morphology: _____

_____

_____

_____

Notes on color and texture: _____

_____

_____

_____

Smell: _____

_____

_____

_____

Other observations: _____

_____

_____

_____

_____

_____

Edible Amanitas: Caesars and Grisettes

Section: *Caesareae* and *Vaginatae*

Eastern Caesar's Amanita
(*Amanita jacksonii*)

Coccora (*Amanita calyptroderma*) egg

Springtime Amanita
(*Amanita velosa*)

More pictures and info here for *Caesareae*:

More pictures and info here for *Vaginatae*:

Taxonomy: *Basidiomycota, Agaricomycetes, Agaricales, Amanitaceae*

Edibility/Toxicity: Great to Incredible Edible (4–5 Stars); edible when raw

Habitat (Association): Grows on ground (Broadleaf/Conifer)

Trophic Mode(s): Ectomycorrhizal

Spores: White (Basidiospore)

Appearance: Distinguished from other, deadly toxic *Amanitas* by a cap, often covered by a large white cottony membrane, easily removed by hand; no small dots or warts on the cap; striations on the edges of the cap; a hollow stem, often filled with white pith; a non-bulbous stem base surrounded by a flaring cottony volva easily peeled off.

Toxic Look-Alikes: Toxic *Amanitas* in section *Amanita* and *Phalloideae*

Safety Notes: Beginners should NEVER even consider eating an *Amanita*. For experienced foragers, learning the robust set of universal characteristics will allow you to safely forage for certain edible species.

Date: _____

Location: _____

_____

_____

_____

Surrounding habitat/context: _____

_____

_____

_____

Morphology: _____

_____

_____

_____

Notes on color and texture: _____

_____

_____

_____

Smell: _____

_____

_____

_____

Other observations: _____

_____

_____

_____

_____

_____

Western Matsutake
(*Tricholoma murrillianum*)

Sandy stem base and cottony partial veil are distinctive characteristics of matsutake.

Eastern Matsutake
(*Tricholoma magnivelare*)

Matsutake (*Tricholoma magnivelare/murrillianum*)

Taxonomy: *Basidiomycota, Agaricomycetes, Agaricales, Tricholomataceae*

Edibility/Toxicity: Great Edible (4 Stars)

Habitat (Association): Grows on ground (Broadleaf/Conifer); best known for growing near pines

Trophic Mode(s): Ectomycorrhizal

Spores: White (Basidiospore)

Appearance: Dense, fibrous (not brittle) flesh; white with reddish-brown scales on cap and stem; a rounded cap with no indent; white gills; a thick cottony veil attached to the stem; a distinctive tough stem with a sandy base buried underground

Notable North American Species: *Tricholoma caligatum, dulciolens, magnivelare, mesoamericanum, murrillianum*

Toxic Look-Alikes: Smith's amanita (*Amanita smithiana*)

Look-Alikes: Imperial cats (*Catathelasma*), short-stemmed russula (*Russula brevipes*)

More pictures and info here:

Date: _____

Location: _____

_____

_____

_____

Surrounding habitat/context: _____

_____

_____

_____

Morphology: _____

_____

_____

_____

Notes on color and texture: _____

_____

_____

_____

Smell: _____

_____

_____

_____

Other observations: _____

_____

_____

_____

_____

_____

Indigo Milkcap
(*Lactarius indigo*)

Saffron Milkcap
(*Lactarius deliciosus*)

Voluminous Milkcap
(*Lactifluus volemus*)

More pictures
and info here:

Milkcaps
(*Lactarius/Lactifluus*)

Taxonomy: *Basidiomycota, Agaricomycetes, Russulales, Russulaceae*

Edibility/Toxicity: Mediocre to Great Edible (2–4 Stars)

Habitat (Association): Grows on ground (Broadleaf/Conifer)

Trophic Mode(s): Ectomycorrhizal

Spores: White to yellowish (Basidiospore)

Appearance: Central stem; no partial veil; a rounded cap with concentric circles, which turns upward in maturity; a chalky, crumbly texture; when immature, exudes a sticky milky latex if cut or injured

Notable North American Species: *Lactarius deliciosus, indigo, paradoxus, rubidus, rubrilacteus; Lactifluus corrugis, hygrophoroides, volemus*

Safety Notes: There is a wide variety of species and some are toxic. Given the diversity of this family, use care when determining edibility. Avoid species that are hairy or fuzzy, have strongly in-rolled caps, or have yellow- or purple-staining latex/milk. Generally, milkcaps that are blue with blue-staining latex (*L. indigo*) or orange with green-staining latex (*L. deliciosus* group) are edible.

Date: _____

Location: _____

_____

_____

_____

Surrounding habitat/context: _____

_____

_____

_____

Morphology: _____

_____

_____

_____

Notes on color and texture: _____

_____

_____

_____

Smell: _____

_____

_____

_____

Other observations: _____

_____

_____

_____

_____

Shrimp Russula
(*Russula xerampelina*)

Green Crackling Russula
(*Russula virescens*)

Short-Stemmed Russula
(*Russula brevipes*)

More pictures
and info here:

Brittlegills (*Russula*)

Taxonomy: *Basidiomycota, Agaricomycetes, Russulales, Russulaceae*

Edibility/Toxicity: Mediocre to Great Edible (2–4 Stars)

Habitat (Association): Grows on ground (Broadleaf/Conifer)

Trophic Mode(s): Ectomycorrhizal

Spores: White to yellowish (Basidiospore)

Appearance: A central stem, no partial veil, a rounded cap that turns upward in maturity, and a chalky, crumbly texture; brittle gills that break easily when handled; visible striations on the edge of the cap and frequently molted or cracked colors on top

Notable North American Species: *Russula brevipes, cyanoxantha, olivacea, parvovirescens, xerampelina*

Safety Notes: There is a wide variety of species and some are toxic. Given the diversity of this family, use care when determining edibility. Avoid specimens with bright red caps—commonly known as sickeners (*R. emetica*)—and any of the blackening or red-staining *Russula* species (*Subnigricantes* section), which are potentially deadly.

Date: _____

Location: _____

_____

_____

_____

Surrounding habitat/context: _____

_____

_____

_____

Morphology: _____

_____

_____

_____

Notes on color and texture: _____

_____

_____

_____

Smell: _____

_____

_____

_____

Other observations: _____

_____

_____

_____

_____

_____

Parasol Mushrooms
(*Macrolepiota*)

Shaggy Parasols
(*Chlorophyllum*)

Fried Chicken Mushrooms
(*Lyophyllum decastes* group)

Velvet Foot / Shanks
(*Flammulina*)

Mushroom Common Name: _____

Taxonomy: _____

Edibility/Toxicity: _____

Habitat (Association): _____

Trophic Mode(s): _____

Spores: _____

Appearance: _____

Notable North American Species: _____

Toxic Look-Alikes: _____

Look-alikes: _____

Date: _____ Location: _____

Surrounding habitat/context: _____

Morphology: _____

Notes on color and texture: _____

Smell: _____

Other observations: _____

Mushroom Common Name: _____

_____

Taxonomy: _____

_____

Edibility/Toxicity: _____

Habitat (Association): _____

_____

Trophic Mode(s): _____

_____

Spores: _____

Appearance: _____

_____

Notable North American Species: _____

_____

Toxic Look-Alikes: _____

Look-alikes: _____

Date: _____ Location: _____

Surrounding habitat/context: _____

_____

Morphology: _____

_____

Notes on color and texture: _____

_____

Smell: _____

_____

Other observations: _____

_____

Mushroom Common Name: _____

_____

Taxonomy: _____

_____

Edibility/Toxicity: _____

Habitat (Association): _____

_____

Trophic Mode(s): _____

_____

Spores: _____

Appearance: _____

_____

Notable North American Species: _____

_____

Toxic Look-Alikes: _____

Look-alikes: _____

Date: _____ Location: _____

Surrounding habitat/context: _____

_____

Morphology: _____

_____

Notes on color and texture: _____

_____

Smell: _____

_____

Other observations: _____

_____

Mushroom Common Name: _____

_____

Taxonomy: _____

_____

Edibility/Toxicity: _____

Habitat (Association): _____

_____

Trophic Mode(s): _____

_____

Spores: _____

Appearance: _____

_____

Notable North American Species: _____

_____

Toxic Look-Alikes: _____

Look-alikes: _____

Date: _____ Location: _____

Surrounding habitat/context: _____

_____

Morphology: _____

_____

Notes on color and texture: _____

_____

Smell: _____

_____

Other observations: _____

_____

CHAPTER 6

Medicinal Mushrooms

Turkey Tail
(*Trametes versicolor*)

Turkey tails on mossy log

Turkey tail pores (top)
compared with smooth
orange *Stereum* (bottom)

More pictures
and info here:

Turkey Tails
(*Trametes versicolor* group)

Taxonomy: *Basidiomycota*, *Agaricomycetes*, *Polyporales*, *Polyporaceae*

Edibility/Toxicity: Medicinal, Special Preparation

Habitat (Association): Grows on wood (Broadleaf)

Trophic Mode(s): Saprotrophic (White Rot)

Spores: White to yellowish (Basidiospore)

Appearance: Wiry, shelf-like clusters with distinctive rings of variable coloration and small hairs on the topside; uniform small white pores on the underside

Look-Alikes: "False Turkey Tail" or Parchment Crusts (*Stereum*), Violet Tooth Polypore (*Trichaptum*), Gilled Polypore (*Trametes betulina*)

Date: _____

Location: _____

_____

_____

_____

Surrounding habitat/context: _____

_____

_____

_____

Morphology: _____

_____

_____

_____

Notes on color and texture: _____

_____

_____

_____

Smell: _____

_____

_____

_____

Other observations: _____

_____

_____

_____

_____

_____

Hemlock Varnish Shelf
(*Ganoderma tsugae*)

Ganoderma polychromum

Pores on the underside
of a reishi

Reishi (*Ganoderma*)

Taxonomy: *Basidiomycota, Agaricomycetes, Polyporales, Ganodermataceae*

Edibility/Toxicity: Medicinal, Special Preparation

Habitat (Association): Grows on wood (Broadleaf/Conifers); occasionally straight out of roots down in the soil

Trophic Mode(s): Saprotrophic (White Rot)

Spores: Brown (Basidiospore)

Appearance: Variable from large woody conk-like shelves to twisted vertical antlers with a distinctive lacquered sheen; often obscured under a heavy cocoa-colored haze of spores

Notable North American Species: *Ganoderma curtisii, oregonense, polychromum, sessile, tsugae, zonatum*

More pictures
and info here:

Date: _____

Location: _____

_____

_____

_____

Surrounding habitat/context: _____

_____

_____

_____

Morphology: _____

_____

_____

_____

Notes on color and texture: _____

_____

_____

_____

Smell: _____

_____

_____

_____

Other observations: _____

_____

_____

_____

_____

Chaga (*Inonotus obliquus*) black sclerotium

Chaga cross section

The chaga fruiting body is a toothy ragged crust.

Chaga (*Inonotus obliquus*)

Taxonomy: *Basidiomycota, Agaricomycetes, Hymenochaetales, Hymenochaetaceae*

Edibility/Toxicity: Medicinal, Special Preparation

Habitat (Association): Grows on wood (Broadleaf), primarily on birch trees in the upper latitudes of the Northern Hemisphere

Trophic Mode(s): Weakly Parasitic, Saprotrophic (White Rot)

Spores: White (Basidiospore)

Appearance: A woody charcoal-like lump. Chaga is not actually a mushroom, but a slow-growing fungal parasite that bursts forth from tree bark as blackened gall-like nutrient storage structure.

Ethics: Given that chaga sclerotia can grow for thirty to eighty years on a host tree, the sustainability of harvesting this fungus has been rightly questioned. If you do choose to harvest chaga, use a tool to cleanly remove it and exercise restraint.

More pictures and info here:

Date: _____

Location: _____

_____

_____

_____

Surrounding habitat/context: _____

_____

_____

_____

Morphology: _____

_____

_____

_____

Notes on color and texture: _____

_____

_____

_____

Smell: _____

_____

_____

_____

Other observations: _____

_____

_____

_____

_____

_____

Cultivated lion's mane

Young lion's mane

Cluster of wild lion's mane
on oak

Lion's Mane
(*Hericium erinaceus*)

Taxonomy: *Basidiomycota, Agaricomycetes, Russulales, Hericiaceae*
Edibility: Incredible Edible (5 Stars)
Habitat (Association): Grows on wood (Broadleaf/Conifer)
Trophic Mode(s): Parasitic, Saprotrophic (White Rot)
Spores: White (Basidiospore)
Appearance: Lion's mane tends to form rounded to lumpy balls with small spines or teeth that elongate as the mushroom matures. Older specimens can turn yellowish.

More pictures
and info here:

Date: _____

Location: _____

_____

_____

_____

Surrounding habitat/context: _____

_____

_____

_____

Morphology: _____

_____

_____

_____

Notes on color and texture: _____

_____

_____

_____

Smell: _____

_____

_____

_____

Other observations: _____

_____

_____

_____

_____

_____

Honorable Mentions

Quinine Conk
(*Laricifomes officinalis*)

Red-Belted Conks
(*Fomitopsis*)

Birch Polypore
(*Fomitopsis betulina*)

Artist's Conks
(*Ganoderma*)

Mushroom Common Name: _____

_____

Taxonomy: _____

_____

Edibility/Toxicity: _____

Habitat (Association): _____

_____

Trophic Mode(s): _____

_____

Spores: _____

Appearance: _____

_____

Notable North American Species: _____

_____

Toxic Look-Alikes: _____

Look-alikes: _____

Date: _____ Location: _____

Surrounding habitat/context: _____

_____

Morphology: _____

_____

Notes on color and texture: _____

_____

Smell: _____

_____

Other observations: _____

_____

Mushroom Common Name: _____

_____

Taxonomy: _____

_____

Edibility/Toxicity: _____

Habitat (Association): _____

_____

Trophic Mode(s): _____

_____

Spores: _____

Appearance: _____

_____

Notable North American Species: _____

_____

Toxic Look-Alikes: _____

Look-alikes: _____

Date: _____ Location: _____

Surrounding habitat/context: _____

_____

Morphology: _____

_____

Notes on color and texture: _____

_____

Smell: _____

_____

Other observations: _____

_____

Mushroom Common Name: _____

_____

Taxonomy: _____

_____

Edibility/Toxicity: _____

Habitat (Association): _____

_____

Trophic Mode(s): _____

_____

Spores: _____

Appearance: _____

_____

Notable North American Species: _____

_____

Toxic Look-Alikes: _____

Look-alikes: _____

Date: _____ Location: _____

Surrounding habitat/context: _____

_____

Morphology: _____

_____

Notes on color and texture: _____

_____

Smell: _____

_____

Other observations: _____

_____

Mushroom Common Name: _____

_____

Taxonomy: _____

_____

Edibility/Toxicity: _____

Habitat (Association): _____

_____

Trophic Mode(s): _____

_____

Spores: _____

Appearance: _____

_____

Notable North American Species: _____

_____

Toxic Look-Alikes: _____

Look-alikes: _____

Date: _____ Location: _____

Surrounding habitat/context: _____

_____

Morphology: _____

_____

Notes on color and texture: _____

_____

Smell: _____

_____

Other observations: _____

_____

CHAPTER 7

Toxic Mushrooms

Deathcap
(*Amanita phalloides*)

Deathcap in hand,
not dangerous to touch

Invasive deathcaps found
growing under oak

Deathcap
(*Amanita phalloides*)

Taxonomy: *Basidiomycota, Agaricomycetes, Agaricales, Amanitaceae*

Edibility/Toxicity: Toxic (Deadly)

Toxin: Amatoxins (Protein/Peptides)

Habitat (Association): Grows on ground (Broadleaf/Conifer); an invasive species throughout North America, particularly in California

Trophic Mode(s): Ectomycorrhizal

Spores: White (Basidiospore)

Appearance: A sickly pale green cap devoid of membrane or any small dots; solid stem with a bulbous base; thick ring around the upper stalk (annulus)

Safety Notes: Deathcaps are responsible for the vast majority (roughly 90 percent) of mushroom-related fatalities worldwide. They are an invasive species from Europe that is actively expanding its range in North America. Please use iNaturalist to help document the spread.

More pictures
and info here:

Date: _____

Location: _____

_____

_____

_____

Surrounding habitat/context: _____

_____

_____

_____

Morphology: _____

_____

_____

_____

Notes on color and texture: _____

_____

_____

_____

Smell: _____

_____

_____

_____

Other observations: _____

_____

_____

_____

_____

Destroying Angel
(*Amanita bisporigera*)

Western Destroying Angel
(*Amanita ocreata*)

Destroying Angels
(*Amanita bisporigera/ ocreata*)

Taxonomy: *Basidiomycota, Agaricomycetes, Agaricales, Amanitaceae*

Edibility/Toxicity: Toxic (Deadly)

Toxin: Amatoxins (Protein/ Peptides)

Habitat (Association): Grows on ground (Broadleaf/Conifer)

Trophic Mode(s): Ectomycorrhizal

Spores: White (Basidiospore)

Appearance: Pure white mushroom usually without warts or membrane on the cap, a well-defined skirt (annulus) on the stem, and a bulbous base; the western species (*Amanita ocreata*) is more solid and chunky than the eastern species; turns slightly ocher and orange on the cap as it ages

Safety Notes: Like the deathcap, these mushrooms are extremely dangerous. No amount of cooking or processing can make them safe to consume.

More pictures
and info here:

Date: _____

Location: _____

_____

_____

_____

Surrounding habitat/context: _____

_____

_____

_____

Morphology: _____

_____

_____

_____

Notes on color and texture: _____

_____

_____

_____

Smell: _____

_____

_____

_____

Other observations: _____

_____

_____

_____

_____

_____

Lorchels (*Gyromitra* and *Paragyromitra*)

Lorchel
(*Gyromitra esculenta*)

Saddle-Shaped Lorchel
(*Paragyromitra infula*)

Lorchel in hand

Alternate Names: "False" Morel, Turban Mushroom

Taxonomy: *Ascomycota, Pezizomycetes, Pezizales, Discinaceae*

Edibility/Toxicity: Toxic (Deadly)

Toxin: Gyromitrin/ Monomethylhydrazine

Habitat (Association): Grows on ground and on wood (Conifer); especially abundant in areas where trees have recently been cut down

Trophic Mode(s): Endophytic/ Saprotrophic (White Rot)

Spores: Whitish to creamy (Ascospore)

Appearance: Funky and alien-looking with brown, wrinkled, lobed caps attaching near the center of the stem; not hollow when cut in half, rather displaying a dense maze of flesh with many folds

Look-Alikes: Elfin saddles (*Helvella*)

Safety Notes: Not to be confused with morels, which often grow in similar habitats. However, lorchels lack morels' conical caps, distinctive pits, and ridges, and are not hollow inside. There is another clade including *G. brunnea, caroliniana, korfii,* and *montana,* which are safe to cook and consume.

More pictures and info here:

Date: _____

Location: _____

_____

_____

_____

Surrounding habitat/context: _____

_____

_____

_____

Morphology: _____

_____

_____

_____

Notes on color and texture: _____

_____

_____

_____

Smell: _____

_____

_____

_____

Other observations: _____

_____

_____

_____

_____

Funeral Bells
(*Galerina marginata* group)

Galerina growing on wood

Moss Bells (*Galerina*)

Galerina badiceps growing on woodchips

Taxonomy: *Basidiomycota, Agaricomycetes, Agaricales, Hymenogastraceae*

Edibility/Toxicity: Toxic (Deadly)

Toxin: Amatoxins (Protein/Peptide)

Habitat (Association): Grows on wood (Broadleaf/Conifer); also found on moss

Trophic Mode(s): Saprotrophic (White Rot)

Spores: Yellowish brown (Basidiospore)

Appearance: Unassuming and little; brown or orange with bell-shaped caps and relatively thin stems that get darker toward the base

Look-Alikes: Velvet Foot / Shanks / enoki (*Flammulina*)

Safety Notes: Because of their innocuous appearance, funeral bells can easily be mistaken by foragers for entheogenic *Psilocybe* species or edible velvet shank / wild enoki mushrooms (*Flammulina*). Pay very close attention to spore color and morphology to avoid a deadly mistake.

More pictures and info here:

Date: _____

Location: _____

_____

_____

_____

Surrounding habitat/context: _____

_____

_____

_____

Morphology: _____

_____

_____

_____

Notes on color and texture: _____

_____

_____

_____

Smell: _____

_____

_____

_____

Other observations: _____

_____

_____

_____

_____

Dapperlings (*Lepiota*)

Dapperling
(*Lepiota castaneidisca*)

Dapperlings in woodchips
(*Lepiota*)

Yellowfoot Dapperling
(*Lepiota magnispora*)

Taxonomy: *Basidiomycota, Agaricomycetes, Agaricales, Agaricaceae*

Edibility/Toxicity: Toxic (Deadly)

Toxin: Amatoxins (Protein/Peptide)

Habitat (Association): Grows on ground (Soil/Duff); common in gardens and urban settings

Trophic Mode(s): Saprotrophic (Composter)

Spores: White (Basidiospore)

Appearance: A cap with a dark center and shaggy scales that radiate outward toward the edges; free gills; a skirt (annulus) that hangs on the stem which is sometimes absent in the smaller *Lepiota* species

Look-Alikes: Parasol mushrooms (*Macrolepiota* and *Chlorophyllum*), shaggy mane inkcap (*Coprinus comatus*)

Safety Notes: As with other amatoxin-containing mushrooms, onset of symptoms can be delayed. Seek treatment immediately if you suspect you have consumed dapperlings, even if no symptoms are yet present.

More pictures
and info here:

Date: _____

Location: _____

_____

_____

_____

Surrounding habitat/context: _____

_____

_____

_____

Morphology: _____

_____

_____

_____

Notes on color and texture: _____

_____

_____

_____

Smell: _____

_____

_____

_____

Other observations: _____

_____

_____

_____

_____

Deadly Webcaps
(*Cortinarius* sect. *Orellani*)

Cortinarius sect. *Orellani*

Deadly Webcaps
(*Cortinarius* section / subgenus *Orellani*)

Alternate Name: Fool's Webcap
Taxonomy: *Basidiomycota, Agaricomycetes, Agaricales, Cortinariaceae*
Edibility/Toxicity: Toxic (Deadly)
Toxin: Orellanine/Cortinarin (Protein)
Habitat (Association): Grows on ground (Conifer)
Trophic Mode(s): Ectomycorrhizal
Spores: Rusty brown (Basidiospore)
Appearance: Medium-sized; coppery brown with dry conical to convex caps covered in fine fibrous scales; pale yellow flesh; a long central stem; a distinctive webby partial veil; yellowish or orangish gills
Safety Notes: This mushroom is particularly dangerous because of the long onset time for symptoms (up to two weeks) and the possibility of permanent and irreversible kidney damage.

More pictures and info here:

Date: _____

Location: _____

_____

_____

_____

Surrounding habitat/context: _____

_____

_____

_____

Morphology: _____

_____

_____

_____

Notes on color and texture: _____

_____

_____

_____

Smell: _____

_____

_____

_____

Other observations: _____

_____

_____

_____

_____

Satan's Bolete
(*Rubroboletus pulcherrimus*)

Satan's bolete
(*Rubroboletus eastwoodiae*)
cross section

Distinctive red pores
under the cap and
reticulation on the stem

More pictures
and info here:

Satan's Boletes
(*Rubroboletus*)

Taxonomy: *Basidiomycota, Agaricomycetes, Boletales, Boletaceae*

Edibility/Toxicity: Toxic (Dangerous)

Toxin: Bolesatine (Protein/Peptide)

Habitat (Association): Grows on ground (Broadleaf/Conifer trees)

Trophic Mode(s): Ectomycorrhizal

Spores: Olive brown (Basidiospore)

Appearance: Sinister-looking with distinctive deep red coloration on the stem and pores; the stem is thick and covered in distinctive webbing (reticulation) near the domed cap and bulbous base; the pores quickly stain dark blue when injured; the flesh also stains dark blue when cut

Look-Alikes: Other red-pored boletes, liver boletes (*Suillellus*), red-mouth boletes (*Boletus subvelutipes*), witches' boletes (*Neoboletus*)

Safety Notes: Consumption can lead to dangerous blood cell clumping. Onset of symptoms occurs within a few hours, and in rare cases can lead to death.

Date: _____

Location: _____

_____

_____

_____

Surrounding habitat/context: _____

_____

_____

_____

Morphology: _____

_____

_____

_____

Notes on color and texture: _____

_____

_____

_____

Smell: _____

_____

_____

_____

Other observations: _____

_____

_____

_____

_____

Western Jack-o'-Lantern
(*Omphalotus olivascens*)

Jack-o'-lantern cluster
in hand

Jack-o'-lantern
(*Omphalotus illudens*) gills

Jack-o'-Lantern Mushrooms
(*Omphalotus*)

Taxonomy: *Basidiomycota, Agaricomycetes, Agaricales, Omphalotaceae*

Edibility/Toxicity: Toxic (Dangerous)

Toxins: Illudins (Sesquiterpene) and Omphalotins

Habitat (Association): Grows on wood (Broadleaf), most often on decaying stumps

Trophic Mode(s): Saprotrophic (White Rot)

Spores: Whitish to yellowish (Basidiospore)

Appearance: Impressively large orange to dingy-orange or green clusters of stem and cap mushrooms which are bioluminescent, occasionally glowing green brightly enough to be seen with the naked eye at night

Look-Alikes: Chanterelles (*Cantharellus*), chicken of the woods (*Laetiporus*), rustgills (*Gymnopilus*)

Safety Notes: Consuming jack-o'-lantern mushrooms will quickly induce severe gastrointestinal distress, though symptoms usually resolve within twelve to twenty-four hours.

More pictures
and info here:

Date: _____

Location: _____

_____

_____

_____

Surrounding habitat/context: _____

_____

_____

_____

Morphology: _____

_____

_____

_____

Notes on color and texture: _____

_____

_____

_____

Smell: _____

_____

_____

_____

Other observations: _____

_____

_____

_____

_____

Sulfur Tufts
(*Hypholoma fasciculare*)

Sulfur Tufts
(*Hypholoma fasciculare*)

Sulfur Tufts growing on
decaying pine

Sulfur tufts under
UV 365 with potassium
hydroxide stain

Taxonomy: *Basidiomycota, Agaricomycetes, Agaricales, Strophariaceae*

Edibility/Toxicity: Toxic (Dangerous)

Toxin: Fasciculols (Steroid-like molecules)

Habitat (Association): Grows on wood (Broadleaf/Conifer)

Trophic Mode(s): Saprotrophic (White Rot)

Spores: Purple-brown (Basidiospore)

Appearance: Brilliantly colorful stem and cap mushrooms, in large clusters, with shades ranging from a vibrant orange to highlighter neon yellow verging onto sickly green

Look-Alikes: Brick top (*Hypholoma lateritium*), magic mushrooms (*Psilocybe*), ringless honey mushrooms (*Desarmillaria*), scalycaps (*Pholiota*), smoky tuft (*Hypholoma capnoides*)

Safety Notes: A wide range of profoundly severe reactions include not only GI symptoms but collapse, paralysis, and impaired vision, though most seem to resolve within a few days. Fortunately, these mushrooms are extremely bitter, which tends to prevent accidental consumption.

More pictures
and info here:

Date: _____

Location: _____

_____

_____

_____

Surrounding habitat/context: _____

_____

_____

_____

Morphology: _____

_____

_____

_____

Notes on color and texture: _____

_____

_____

_____

Smell: _____

_____

_____

_____

Other observations: _____

_____

_____

_____

_____

Fool's Funnel
(*Clitocybe rivulosa*)

Fool's funnel
seen growing in grass

Fool's Funnel (*Clitocybe / Collybia rivulosa*)

Taxonomy: *Basidiomycota*, *Agaricomycetes*, *Agaricales*, *Clitocybaceae*

Edibility/Toxicity: Toxic (Dangerous)

Toxin: Muscarine (Neurotoxin)

Habitat (Association): Grows on ground (Grass), particularly on lawns and sandy dunes

Trophic Mode(s): Saprotrophic (White Rot)

Spores: White (Basidiospore)

Appearance: A cap with in-rolled edges when young; flattens out in maturity, developing a small dent in the center and concentric rings with pinkish to brown discoloration

Look-alikes: Other species of funnels, the spy (*Clitopilus prunulus*)

Safety Notes: Human deaths due to consuming fool's funnel are rare, but they are especially hazardous for dogs because they contain muscarine and are common in grassy areas. Veterinarians can administer atropine as an antidote but be sure to keep a sample of the mushroom to assist with diagnosis.

More pictures
and info here:

Date: _____

Location: _____

_____

_____

_____

Surrounding habitat/context: _____

_____

_____

_____

Morphology: _____

_____

_____

_____

Notes on color and texture: _____

_____

_____

_____

Smell: _____

_____

_____

_____

Other observations: _____

_____

_____

_____

_____

_____

Yellow Staining Agaric
(*Agaricus xanthodermus*)

Big cluster of yellow
staining agarics

Yellow staining agarics
on grass

More pictures
and info here:

Yellow Staining Agarics
(*Agaricus* section
Xanthodermatei)

Taxonomy: *Basidiomycota,
Agaricomycetes, Agaricales,
Agaricaceae*
Edibility/Toxicity: Toxic (Caution)
Toxin: Phenol (GI upset)
Habitat (Association): Grows on
ground (Soil/Grass); particularly
common on lawns and in urban
areas
Trophic Mode(s): Saprotrophic
(Composter)
Spores: Brown (Basidiospore)
Appearance: White to brownish
coloration on the cap; remnants
of a partial veil on the stem;
crowded free gills that go from
pink to brown with maturity;
will stain yellow when handled,
particularly at the base of the
stem and on the edges of the
cap
Look-Alikes: Field mushroom
(*A. campestris*), horse mushroom
(*A. arvensis*), almond-scented
agarics (*Agaricus*)
Safety Notes: Can cause severe
GI upset. Yellow staining
agarics look very similar to
a number of edible *Agaricus*
species, but their yellow color
and an unpleasant smell (like
disinfectant, ink, or paste) are
most noticeable when heated.

Date: _____

Location: _____

_____

_____

_____

Surrounding habitat/context: _____

_____

_____

_____

Morphology: _____

_____

_____

_____

Notes on color and texture: _____

_____

_____

_____

Smell: _____

_____

_____

_____

Other observations: _____

_____

_____

_____

_____

Fly Agarics
(*Amanita muscaria* group)

Amanita muscaria
var. *flavivolvata*

Fly Agaric
(*Amanita muscaria*)

Fly agaric
(*Amanita muscaria*) button

Taxonomy: *Basidiomycota, Agaricomycetes, Agaricales, Amanitaceae*

Edibility/Toxicity: Toxic; edible with special preparation

Toxin: Ibotenic Acid / Muscimol (Isoxazole derivatives)

Habitat (Association): Grows on ground (Broadleaf/Conifers)

Trophic Mode(s): Ectomycorrhizal

Spores: White (Basidiospore)

Appearance: Highly recognizable; red or orange cap with white or yellow warts; whitish stem with a skirt and distinctively bulbous base

Safety Notes: While fly agaric can be poisonous—and induce sleepiness, GI upset, and entheogenic effects in a dose-dependent manner—they do not contain life-threatening amatoxins, making them far less dangerous than many other *Amanitas*. These mushrooms can even be prepared as food by boiling in clean water twice to remove toxins, then cooking as usual.

More pictures
and info here:

Date: _____

Location: _____

_____

_____

_____

Surrounding habitat/context: _____

_____

_____

_____

Morphology: _____

_____

_____

_____

Notes on color and texture: _____

_____

_____

_____

Smell: _____

_____

_____

_____

Other observations: _____

_____

_____

_____

_____

_____

Green-spored parasols
(*Chlorophyllum molybdites*)
on grass

Green-Spored Parasol
(*Chlorophyllum molybdites*)

Green-Spored Parasol
(*Chlorophyllum molybdites*)

More pictures
and info here:

Green-Spored Parasol
(*Chlorophyllum molybdites*)

Taxonomy: *Basidiomycota, Agaricomycetes, Agaricales, Agaricaceae*

Edibility/Toxicity: Toxic (Caution)

Toxin: Molybdophyllysin (Protein)

Habitat (Association): Grows on ground (Grass)

Trophic Mode(s): Saprotrophic (Composter)

Spores: Green (Basidiospore)

Appearance: Brown scales atop a white cap; a softly colored brownish and whitish stem with a well-defined collar-like ring around it; initially white gills that quickly develop a greenish tinge as they mature

Look-Alikes: Parasols (*Macrolepiota* and white-spored *Chlorophyllum*), shaggy mane inkcap (*Coprinus comatus*)

Safety Notes: Similar in appearance to edible parasol mushrooms, this is one of the most commonly consumed poisonous mushrooms in North America. This species makes so many people so ill that it has earned the nickname "the vomiter." While they don't usually lead to more serious health complications for humans, they can be deadly for dogs.

Date: _____

Location: _____

_____

_____

_____

Surrounding habitat/context: _____

_____

_____

_____

Morphology: _____

_____

_____

_____

Notes on color and texture: _____

_____

_____

_____

Smell: _____

_____

_____

_____

Other observations: _____

_____

_____

_____

_____

_____

Earthballs (*Scleroderma*)

Sometimes called
Pigskin Puffballs given
the warty outer exterior

Initially whitish inside,
the gleba turns purple to
black as it matures.

Earthballs (*Scleroderma*)

Taxonomy: *Basidiomycota,*
Agaricomycetes, Boletales,
Sclerodermataceae

Edibility/Toxicity: Toxic (Caution)

Toxins: Sclerocitrin (Pigment),
vulpinic acid

Habitat (Association): Grows on
ground (Broadleaf/Conifer);
a particularly important
mycorrhizal partner for trees

Trophic Mode(s): Ectomycorrhizal

Spores: Gray to purple-brown to
purple-black (Basidiospore)

Appearance: Unassuming
yellowish or tan terrestrial
lumps with a rough or cracked
to warty exterior and a dense
mass of spores inside

Safety notes: Consumption can
cause GI upset and neurological
symptoms. Exposure to the
spores can cause symptoms
similar to allergies: acute stuffy
and runny nose, tears, and eye
inflammation. Earthballs are
particularly toxic to dogs.

More pictures
and info here:

Date: _____

Location: _____

_____

_____

_____

Surrounding habitat/context: _____

_____

_____

_____

Morphology: _____

_____

_____

_____

Notes on color and texture: _____

_____

_____

_____

Smell: _____

_____

_____

_____

Other observations: _____

_____

_____

_____

_____

_____

Poison Pies
(*Hebeloma*)

Scalycaps
(*Pholiota squarrosa* group)

Scaly "Chanterelle"
(*Turbinellus floccosus*)

Cinnamon Bracket
(*Hapalopilus rutilans*)

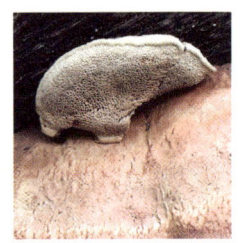

Mushroom Common Name: _____

_____

Taxonomy: _____

_____

Edibility/Toxicity: _____

Habitat (Association): _____

_____

Trophic Mode(s): _____

_____

Spores: _____

Appearance: _____

_____

Notable North American Species: _____

_____

Toxic Look-Alikes: _____

Look-alikes: _____

Date: _____ Location: _____

Surrounding habitat/context: _____

_____

Morphology: _____

_____

Notes on color and texture: _____

_____

Smell: _____

_____

Other observations: _____

_____

Mushroom Common Name: _____

_____

Taxonomy: _____

_____

Edibility/Toxicity: _____

Habitat (Association): _____

_____

Trophic Mode(s): _____

_____

Spores: _____

Appearance: _____

_____

Notable North American Species: _____

_____

Toxic Look-Alikes: _____

Look-alikes: _____

Date: _____ Location: _____

Surrounding habitat/context: _____

_____

Morphology: _____

_____

Notes on color and texture: _____

_____

Smell: _____

_____

Other observations: _____

_____

Mushroom Common Name: _____

Taxonomy: _____

Edibility/Toxicity: _____

Habitat (Association): _____

Trophic Mode(s): _____

Spores: _____

Appearance: _____

Notable North American Species: _____

Toxic Look-Alikes: _____

Look-alikes: _____

Date: _____ Location: _____

Surrounding habitat/context: _____

Morphology: _____

Notes on color and texture: _____

Smell: _____

Other observations: _____

Mushroom Common Name: _____

_____

Taxonomy: _____

_____

Edibility/Toxicity: _____

Habitat (Association): _____

_____

Trophic Mode(s): _____

_____

Spores: _____

Appearance: _____

_____

Notable North American Species: _____

_____

Toxic Look-Alikes: _____

Look-alikes: _____

Date: _____ Location: _____

Surrounding habitat/context: _____

_____

Morphology: _____

_____

Notes on color and texture: _____

_____

Smell: _____

_____

Other observations: _____

_____

CHAPTER 8

Dye Mushrooms

Dyer's Polypore
(*Phaeolus schweinitzii*)

Pores on the underside of
a Dyer's polypore

Mikhael Crystallah-Selk
holding a mature
Dyer's polypore

Dyer's Polypores
(*Phaeolus schweinitzii* group)

Taxonomy: *Basidiomycota, Agaricomycetes, Polyporales, Phaeolaceae*

Edibility/Toxicity: Inedible (Potentially Toxic)

Habitat (Association): Grows on wood (Conifer)

Trophic Mode(s): Parasitic, Saprotrophic (Brown Rot)

Spores: White to yellowish (Basidiospore)

Appearance: Large, initially soft, yellow, fuzzy rosettes and layered shelves; become dark brown and woody as they mature.

Dye Notes: Depending on pH and mordant used, Dyer's polypore can produce a range of green to yellow to brown pigments.

More pictures
and info here:

Date: _____

Location: _____

_____

_____

_____

Surrounding habitat/context: _____

_____

_____

_____

Morphology: _____

_____

_____

_____

Notes on color and texture: _____

_____

_____

_____

Smell: _____

_____

_____

_____

Other observations: _____

_____

_____

_____

_____

_____

Dyer's Corts
(*Cortinarius* section / subgenus *Dermocybe*)

Cortinarius smithii

Cortinarius cinnamomeus

Taxonomy: *Basidiomycota, Agaricomycetes, Agaricales, Cortinariaceae*

Edibility/Toxicity: Inedible (Potentially Toxic)

Habitat (Association): Grows on ground (Conifer)

Trophic Mode(s): Ectomycorrhizal

Spores: Rusty brown (Basidiospore)

Appearance: A dry cap; relatively thin, dry stem; bright colors (especially on the gills) ranging from yellow to orange to red; distinctive webby partial veil that covers the gills when immature.

Dye Notes: Depending on pH and mordant used, these *Cortinarius* can produce a range of brilliant pink, red, purple to yellow and orange pigments.

Pile of Dyer's Cort growing near pine

More pictures and info here:

Date: _____

Location: _____

_____

_____

_____

Surrounding habitat/context: _____

_____

_____

_____

Morphology: _____

_____

_____

_____

Notes on color and texture: _____

_____

_____

_____

Smell: _____

_____

_____

_____

Other observations: _____

_____

_____

_____

_____

_____

Bitter Tooth (*Hydnellum*)

Taxonomy: *Basidiomycota*, *Agaricomycetes*, *Thelephorales*, *Bankeraceae*

Edibility/Toxicity: Nontoxic (1 Star); tough and bitter

Habitat (Association): Grows on ground (Broadleaf/Conifer)

Trophic Mode(s): Ectomycorrhizal

Spores: Brown (Basidiospore)

Appearance: Tough stemmed mushrooms that grow from small nubs into spreading circular zonate shields; when young, the velvety growing edges can be covered in droplets of guttation; as the cap matures, they darken and harden; underside of the cap is covered in small teeth; will often stain or discolor when handled.

Dye Notes: Depending on pH and mordant used, *Hydnellum* can produce a range of blue-green to gray pigments.

Bloody droplets of guttations on Bleeding Tooth (*Hydnellum peckii*)

Guttation on growing edge of Blue Tooth (*Hydnellum cyanopodium*)

Spines on underside of Orange Tooth (*Hydnellum aurantiacum*)

More pictures and info here:

Date: _____

Location: _____

_____

_____

_____

Surrounding habitat/context: _____

_____

_____

_____

Morphology: _____

_____

_____

_____

Notes on color and texture: _____

_____

_____

_____

Smell: _____

_____

_____

_____

Other observations: _____

_____

_____

_____

_____

_____

Dyeball (*Pisolithus*) clump

Dyeball (*Pisolithus*)
cross section

Close-up view
of dyeball peridioles

Dead Man's Foot
(*Pisolithus*)

Alternate Names: Dye Balls, Dog Turd Fungus

Taxonomy: *Basidiomycota, Agaricomycetes, Boletales, Sclerodermataceae*

Edibility/Toxicity: Nontoxic (1 Star)

Habitat (Association): Grows on ground (Broadleaf/Conifer)

Trophic Mode(s): Ectomycorrhizal

Spores: Ocher to Brown (Basidiospore)

Appearance: Tough dingy whitish, blackish, and dusty brown lumps and haphazard piles reminiscent of dried dog feces; when cut open reveals shockingly beautiful ombrés of bright white, yellow, and rust against black backgrounds; depending on pH and mordant used, dye balls produce a range of tan to brown pigments

Look-Alikes: Earthballs (*Scleroderma*), truffle-like mushrooms when immature

More pictures
and info here:

Date: _____

Location: _____

_____

_____

_____

Surrounding habitat/context: _____

_____

_____

_____

Morphology: _____

_____

_____

_____

Notes on color and texture: _____

_____

_____

_____

Smell: _____

_____

_____

_____

Other observations: _____

_____

_____

_____

_____

_____

CHAPTER 9

Gasteroid Mushrooms

Bridal Veil Stinkhorn
(*Phallus indusiatus*)

Devil's Fingers
(*Clathrus archeri*)

Devil's Dipstick
(*Mutinus elegans*)

More pictures
and info here:

Stinkhorns (*Phallaceae*)

Taxonomy: *Basidiomycota, Agaricomycetes, Phallales, Phallaceae/Clathraceae/ Lysuraceae*

Edibility/Toxicity: Mediocre/Fair (2 Stars)

Habitat (Association): Grows on ground (Mulch/Soil), often in woodchips and moist areas with decaying organic material

Trophic Mode(s): Saprotrophic (White Rot)

Spores: Stinky olive-brown to brown goo (Basidiospore)

Appearance: Immature rubbery white eggs explode into various fantastical, bizarre, and spectacularly alien-looking shapes adorned with a gooey mass of sticky spores.

Look-Alikes: *Amanita* egg when immature, puffballs (*Lycoperdaceae*)

Date: _____

Location: _____
_____
_____
_____

Surrounding habitat/context: _____
_____
_____
_____

Morphology: _____
_____
_____
_____

Notes on color and texture: _____
_____
_____
_____

Smell: _____
_____
_____
_____

Other observations: _____
_____
_____
_____
_____
_____

Earthstars (*Geastraceae*)

Taxonomy: *Basidiomycota, Agaricomycetes, Geastrales, Geastraceae*

Edibility/Toxicity: Nontoxic (1 Star); inedible and tough

Habitat (Association): Grows on ground (Mulch/Grass); most common in forests but also occurring in woodchips and urban environments

Trophic Mode(s): Saprotrophic (White Rot)

Spores: Brown (Basidiospore)

Appearance: Diminutive rounded balls open at the top with distinctive starlike petals

Look-Alikes: Hygroscopic earthstars (*Astraeus*)

Earthstar (*Geastrum*)

Earthstar growing in duff

Hygroscopic Earthstars (*Astraeus*) look similar but are unrelated, belonging to the *Boletales* order.

More pictures and info here:

Date: _____

Location: _____

_____

_____

_____

Surrounding habitat/context: _____

_____

_____

_____

Morphology: _____

_____

_____

_____

Notes on color and texture: _____

_____

_____

_____

Smell: _____

_____

_____

_____

Other observations: _____

_____

_____

_____

_____

_____

Bird's Nest Fungi
(*Nidulariaceae*)

Common Bird's Nest
(*Crucibulum*)

Bird's Nest Fungi (*Cyathus*)

Woolly Bird's Nest Fungus
(*Nidula niveotomentosa*)

Alternate Name: Splash Cups
Taxonomy: *Basidiomycota, Agaricomycetes, Agaricales, Nidulariaceae*
Edibility/Toxicity: Nontoxic (1 Star); inedible and tough
Habitat (Association): Grows on wood (Mulch/Sticks), especially in landscaped woodchip and mulch beds
Trophic Mode(s): Saprotrophic (White Rot)
Spores: Hidden inside egg-like peridioles (Basidiospore)
Appearance: Adorable little "nests" or cups covered in a gelatinous membrane that wears away when it rains to reveal tiny egg-like structures (peridioles) only the size of a lentil, which in turn get splashed out of the cup by raindrops to spread spores

More pictures and info here:

Date: _____

Location: _____

_____

_____

_____

Surrounding habitat/context: _____

_____

_____

_____

Morphology: _____

_____

_____

_____

Notes on color and texture: _____

_____

_____

_____

Smell: _____

_____

_____

_____

Other observations: _____

_____

_____

_____

_____

Cannonball Fungus
(*Sphaerobolus stellatus*)

Close-up of
cannonball or peridiole

Cannonball Fungi
(*Sphaerobolus*)

Taxonomy: *Basidiomycota, Agaricomycetes, Geastrales, Geastraceae*

Edibility/Toxicity: Nontoxic (1 Star); inedible and tough

Habitat (Association): Grows on ground (Mulch), especially in landscaped woodchip and mulch beds

Trophic Mode(s): Saprotrophic (White Rot)

Spores: White round peridiole (Basidiospore)

Appearance: Small orange balls that explode open, launching a small white "cannonball" of spores up to ten to fifteen feet away, leaving a small opened hole with starlike petals

More pictures
and info here:

Date: _____

Location: _____

_____

_____

_____

Surrounding habitat/context: _____

_____

_____

_____

Morphology: _____

_____

_____

_____

Notes on color and texture: _____

_____

_____

_____

Smell: _____

_____

_____

_____

Other observations: _____

_____

_____

_____

_____

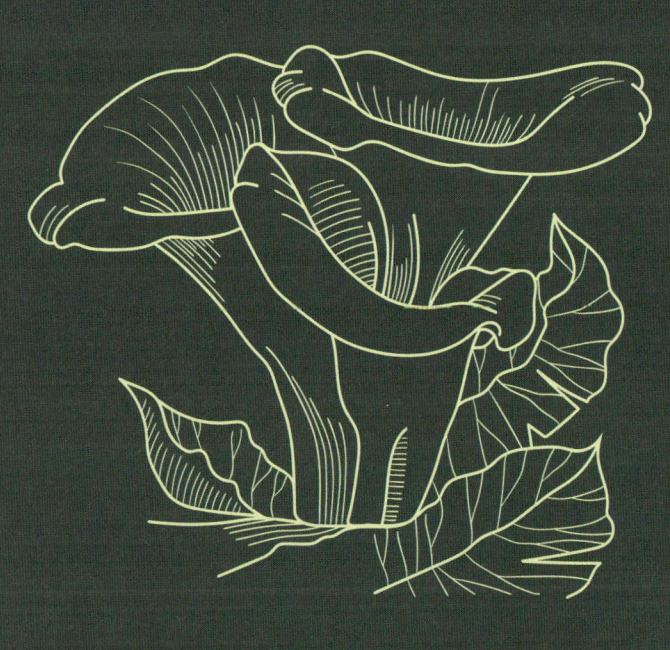

CHAPTER 10

Fungal Oddities

Ramaria rainbow

Green Ramaria
(*Ramaria apiculata*)

Violet Coral
(*Clavaria zollingeri*)

More pictures and info here for *Ramaria*:

More pictures and info here for *Clavariaceae*:

Corals, Antlers, and Spindles
(*Ramaria* and *Clavariaceae*)

Taxonomy: *Basidiomycota, Agaricomycetes, Gomphales, Gomphaceae*

Edibility/Toxicity: Mixed edibility; some cause GI upset while others are Good Edibles (3 Stars)

Habitat (Association): Grows on ground (Broadleaf/Conifer)

Trophic Mode(s): Ectomycorrhizal

Spores: Yellowish to orangish (Basidiospore)

Appearance: Branching bunches, reminiscent of underwater corals; color ranging from drab shades of cream, beige, and brown to brilliant yellows, oranges, reds, greens, and purples

Safety Notes: The edibility of corals is variable. In North America, they are generally avoided except for a few specific species, as some are known to cause fairly severe GI upset. All can have a laxative effect if too much is consumed.

Date: _____

Location: _____

_____

_____

_____

Surrounding habitat/context: _____

_____

_____

_____

Morphology: _____

_____

_____

_____

Notes on color and texture: _____

_____

_____

_____

Smell: _____

_____

_____

_____

Other observations: _____

_____

_____

_____

_____

Witches' Butter (*Tremella*)

Golden Ear
(*Naematelia aurantia*)
parasitizing *Stereum*

Leafy Brain
(*Phaeotremella foliacea*)

Jelly Mushrooms
(*Tremellaceae*)

Taxonomy: *Basidiomycota, Tremellomycetes, Tremellales, Tremellaceae*

Edibility/Toxicity: Mediocre (2 Star) to Good Edible (3 Star); Medicinal

Habitat (Association): Grows on wood (Broadleaf)

Trophic Mode(s): Parasitic

Spores: White (Basidiospore)

Appearance: Indeterminate, folded, brain-like blobs and lobes—brown, black, yellow, white, or dingy cream—can shrivel up in dry conditions, then rehydrate to their gelatinous, jellylike texture when moisture returns

Look-Alikes: Jelly spots (*Dacrymyces*), which grow on conifers

More pictures
and info here:

Date: _____

Location: _____

_____

_____

_____

Surrounding habitat/context: _____

_____

_____

_____

Morphology: _____

_____

_____

_____

Notes on color and texture: _____

_____

_____

_____

Smell: _____

_____

_____

_____

Other observations: _____

_____

_____

_____

_____

_____

Cordyceps militaris
on larvae

Cordyceps tenuipes

*Ophiocordyceps
dipterigena*

More pictures
and info here for
Cordycipitaceae:

More pictures
and info here for
Ophiocordycipiticeae:

Cordyceps
(*Cordycipitaceae* and *Ophiocordycipitaceae*)

Taxonomy: *Ascomycota, Sordariomycetes, Hypocreales, Cordycipitaceae*

Edibility/Toxicity: Nontoxic (1 Star)

Habitat (Association): Grows on ground (Soil/Duff), from the bodies of dead, parasitized insects (sometimes buried deep underground), in mossy sites, forest, or grassland

Trophic Mode(s): Parasitic

Spores: White (Ascospore)

Appearance: Small stalks growing directly out of insects or larvae; range in color from red-orange to yellow to white to brown; some species are covered with tiny pimple-like dots (perithecia).

Look-Alikes: Truffleclubs (*Tolypocladium*) and other entomopathogenic fungi

Safety Notes: Cultivated *Cordyceps militaris* are delicious when cooked. *Ophiocordyceps sinensis* are used medicinally. Besides these, it's wise to avoid consuming unknown species.

Date: _____

Location: _____

_____

_____

_____

Surrounding habitat/context: _____

_____

_____

_____

Morphology: _____

_____

_____

_____

Notes on color and texture: _____

_____

_____

_____

Smell: _____

_____

_____

_____

Other observations: _____

_____

_____

_____

_____

_____

Candlesnuff Fungus
(*Xylaria hypoxylon*)

Dead Man's Fingers
(*Xylaria polymorpha*)

King Alfred's Cakes
(*Daldinia concentrica*)

Carbon Fungi
(*Xylariales*)

Taxonomy: *Ascomycota, Sordariomycetes, Xylariales, Xylariaceae*

Edibility/Toxicity: Nontoxic (1 Star); inedible and tough

Habitat (Association): Grows on wood (Broadleaf/Conifer)

Trophic Mode(s): Saprotrophic (Soft Rot)

Spores: White asexual spores (conidia), dark brown to black sexual spores (Ascospore)

Appearance: Small, tough, hard, and long-lasting structures in various shapes including stalks, antlers, crusts, balls, and cushions. Initially they start out whitish, producing asexual spores (conidia), then turn hard and black, slowly releasing spores from small pits on the bumpy surface (perithecia).

More pictures and info here:

Date: _____

Location: _____

_____

_____

_____

Surrounding habitat/context: _____

_____

_____

_____

Morphology: _____

_____

_____

_____

Notes on color and texture: _____

_____

_____

_____

Smell: _____

_____

_____

_____

Other observations: _____

_____

_____

_____

_____

_____

Rusts *(Pucciniales)*

Taxonomy: *Basidiomycota, Pucciniomycetes, Pucciniales*
Edibility/Toxicity: Inedible (Potentially Toxic)
Habitat (Association): Plant parasite often growing on leaves, stalks, flowers, and seed pods
Trophic Mode(s): Parasitic
Spores: Orange (Basidiospore)
Appearance: Rusts are highly variable, switching between plant hosts during their life cycle and producing multiple kinds of spores; the most visible sign of rust infection is a dusty orange crust (pycnia) that forms on infected plants.

Cedar Apple Rust
(*Gymnosporangium juniperi-virginianae*)

Rust pycnia
(*Uromyces aureus*)

Blackberry Orange Rust
(*Gymnoconia peckiana*)

More pictures and info here:

Date: _____

Location: _____

_____

_____

_____

Surrounding habitat/context: _____

_____

_____

_____

Morphology: _____

_____

_____

_____

Notes on color and texture: _____

_____

_____

_____

Smell: _____

_____

_____

_____

Other observations: _____

_____

_____

_____

_____

_____

Smut (*Ustilago*)

Brome Smut Fungus
(*Ustilago bullata*)

Huitlacoche, or Corn Smut
(*Mycosarcoma maydis*)

More pictures
and info here:

Smuts (*Ustilaginaceae*)

Taxonomy: *Basidiomycota, Ustilaginomycetes, Ustilaginales, Ustilaginaceae*

Edibility/Toxicity: Edibility not recommended

Habitat (Association): Plant parasite that infects cereal grains

Trophic Mode(s): Parasitic

Spores: Black (Basidiospore)

Appearance: Variable depending on the host, generally forming black swollen galls on grains, releasing a sooty puff of spores when mature.

Safety Notes: While most smuts are potentially toxic, corn smut or huitlacoche is a traditional (and delicious) food consumed in Mexico.

Date: _____

Location: _____

_____

_____

_____

Surrounding habitat/context: _____

_____

_____

_____

Morphology: _____

_____

_____

_____

Notes on color and texture: _____

_____

_____

_____

Smell: _____

_____

_____

_____

Other observations: _____

_____

_____

_____

_____

CHAPTER 11

Ascomycete Cup Fungi

Scarlet Elf Cups
(Sarcoscypha)

Orange Peel Fungus
(Aleuria aurantia)

Green Elf Cups
(Chlorociboria)

Yellow Fairy Cups
(Calycina)

Eyelash Cup
(Scutellinia scutellata)

Phoenix Cup
(Peziza violacea)

More pictures
and info here for
Colorful Cups:

More pictures
and info here for
Colorful Discs:

Colorful Asco Cup Fungi
(*Aleuria, Ascocoryne, Calcyclina, Chlorociboria, Peziza, Sarcoscypha, Scutellinia, Sowerbyella*)

Taxonomy: *Ascomycota, Pezizomycetes, Pezizales, Helotiales*

Edibility: Not Recommended

Habitat (Association): Grows on wood, on ground (variable)

Trophic Mode(s): Parasitic, Saprotrophic, Ectomycorrhizal

Spores: White to whitish (Ascospore)

Appearance: Ascomycete cup fungi are wildly variable in their appearance, but generally they will form cups (with or without stalks) or more compressed disc-like structures directly on wood or amid decaying organic material.

Safety Notes: While most Asco cups are not desirable edibles, the scarlet elf cups can be pickled and used in salads.

Date: _____

Location: _____

_____

_____

_____

Surrounding habitat/context: _____

_____

_____

_____

Morphology: _____

_____

_____

_____

Notes on color and texture: _____

_____

_____

_____

Smell: _____

_____

_____

_____

Other observations: _____

_____

_____

_____

_____

Date: _____

Location: _____

_____

_____

_____

Surrounding habitat/context: _____

_____

_____

_____

Morphology: _____

_____

_____

_____

Notes on color and texture: _____

_____

_____

_____

Smell: _____

_____

_____

_____

Other observations: _____

_____

_____

_____

_____

_____

Date: _____

Location: _____

_____

_____

_____

Surrounding habitat/context: _____

_____

_____

_____

Morphology: _____

_____

_____

_____

Notes on color and texture: _____

_____

_____

_____

Smell: _____

_____

_____

_____

Other observations: _____

_____

_____

_____

_____

Date: _____

Location: _____

_____

_____

_____

Surrounding habitat/context: _____

_____

_____

_____

Morphology: _____

_____

_____

_____

Notes on color and texture: _____

_____

_____

_____

Smell: _____

_____

_____

_____

Other observations: _____

_____

_____

_____

_____

_____

Date: _____

Location: _____

_____

_____

_____

Surrounding habitat/context: _____

_____

_____

_____

Morphology: _____

_____

_____

_____

Notes on color and texture: _____

_____

_____

_____

Smell: _____

_____

_____

_____

Other observations: _____

_____

_____

_____

_____

Date: _____

Location: _____

_____

_____

_____

Surrounding habitat/context: _____

_____

_____

_____

Morphology: _____

_____

_____

_____

Notes on color and texture: _____

_____

_____

_____

Smell: _____

_____

_____

_____

Other observations: _____

_____

_____

_____

_____

CHAPTER 12

Honorable Mentions: Lichens

Honorable Mentions

Spindles and Structured Lichens (*Cladonia/Cladina*)
Alternate Names:
Lipstick Lichen,
Soldier Lichen,
Pixie Cup Lichen,
Reindeer Lichen

Beard Lichens
(*Usnea*)
Alternate Name:
Old Man's Beard

Sunburst Lichens
(*Xanthoria*)

Map Lichens
(*Rhizocarpon*)

Shield Lichens
(*Parmelia*)

Wolf Lichens
(*Letharia*)

Mushroom Common Name: _____

_____

Taxonomy: _____

_____

Edibility/Toxicity: _____

Habitat (Association): _____

_____

Trophic Mode(s): _____

_____

Spores: _____

Appearance: _____

_____

Notable North American Species: _____

_____

Toxic Look-Alikes: _____

Look-alikes: _____

Date: _____ Location: _____

Surrounding habitat/context: _____

_____

Morphology: _____

_____

Notes on color and texture: _____

_____

Smell: _____

_____

Other observations: _____

_____

Mushroom Common Name: _____

Taxonomy: _____

Edibility/Toxicity: _____

Habitat (Association): _____

Trophic Mode(s): _____

Spores: _____

Appearance: _____

Notable North American Species: _____

Toxic Look-Alikes: _____

Look-alikes: _____

Date: _____ Location: _____

Surrounding habitat/context: _____

Morphology: _____

Notes on color and texture: _____

Smell: _____

Other observations: _____

Mushroom Common Name: _____

_____

Taxonomy: _____

_____

Edibility/Toxicity: _____

Habitat (Association): _____

_____

Trophic Mode(s): _____

_____

Spores: _____

Appearance: _____

_____

Notable North American Species: _____

_____

Toxic Look-Alikes: _____

Look-alikes: _____

Date: _____ Location: _____

Surrounding habitat/context: _____

_____

Morphology: _____

_____

Notes on color and texture: _____

_____

Smell: _____

_____

Other observations: _____

_____

Mushroom Common Name: _____

_____

Taxonomy: _____

_____

Edibility/Toxicity: _____

Habitat (Association): _____

_____

Trophic Mode(s): _____

_____

Spores: _____

Appearance: _____

_____

Notable North American Species: _____

_____

Toxic Look-Alikes: _____

Look-alikes: _____

Date: _____ Location: _____

Surrounding habitat/context: _____

_____

Morphology: _____

_____

Notes on color and texture: _____

_____

Smell: _____

_____

Other observations: _____

_____

Mushroom Common Name: _____

_____

Taxonomy: _____

_____

Edibility/Toxicity: _____

Habitat (Association): _____

_____

Trophic Mode(s): _____

_____

Spores: _____

Appearance: _____

_____

Notable North American Species: _____

_____

Toxic Look-Alikes: _____

Look-alikes: _____

Date: _____ Location: _____

Surrounding habitat/context: _____

_____

Morphology: _____

_____

Notes on color and texture: _____

_____

Smell: _____

_____

Other observations: _____

_____

Mushroom Common Name: _____

Taxonomy: _____

Edibility/Toxicity: _____

Habitat (Association): _____

Trophic Mode(s): _____

Spores: _____

Appearance: _____

Notable North American Species: _____

Toxic Look-Alikes: _____

Look-alikes: _____

Date: _____ Location: _____

Surrounding habitat/context: _____

Morphology: _____

Notes on color and texture: _____

Smell: _____

Other observations: _____

CHAPTER 13

Not Fungi but Fascinating

Slime mold plasmodium

Wolf's Milk Slime Mold
(*Lycogala*)

Chocolate Tube
Slime Mold (*Stemonitis*)

More pictures
and info here:

Slime Molds
(*Myxogastria*)

Taxonomy: *Eukaryota*, *Amoebozoa*, *Conosa*, *Eumycetozoa*
Edibility/Toxicity: Inedible
Habitat (Association): Grows on wood and on ground (Soil/Duff), particularly rotting wood, forest floors, lawns, woodchips, and garden mulch
Trophic Mode(s): Saprotrophic
Spores: Black to brown
Appearance: Not a mold, nor even a fungus, but a single-cell amoeba-like organism with a multiphase life cycle. Not visible to the naked eye in the first phase, in the second phase it's a brightly colored amorphous goo (plasmodium), and in the final phase it forms distinctive, exotic-looking, spore-bearing fruiting bodies (sporangia).

Date: _____

Location: _____

_____

_____

_____

Surrounding habitat/context: _____

_____

_____

_____

Morphology: _____

_____

_____

_____

Notes on color and texture: _____

_____

_____

_____

Smell: _____

_____

_____

_____

Other observations: _____

_____

_____

_____

_____

_____

Ghost Pipes
(*Monotropa uniflora*)

Gnome Plant
(*Hemitomes congestum*)

Snow Plant
(*Sarcodes sanguinea*)

Mycoheterotrophic Plants (*Ericaceae*)

Taxonomy: *Asterids, Ericales, Ericaceae, Monotropoideae*

Edibility/Toxicity: Inedible (Potentially Toxic)

Habitat (Association): Grows on ground (Soil/Duff), bursting straight from the soil in the forest understory

Trophic Mode(s): Parasitic, Symbiotic

Spores: N/A

Appearance: Plants that have no green leaves and do not photosynthesize, instead tapping into the mycelial networks of fungi to survive; usually hidden from view, living only underground as fleshy stem roots (rhizomes); they annually produce eye-catching clusters of flowers on stalks.

More pictures and info here:

Date: _____

Location: _____

_____

_____

_____

Surrounding habitat/context: _____

_____

_____

_____

Morphology: _____

_____

_____

_____

Notes on color and texture: _____

_____

_____

_____

Smell: _____

_____

_____

_____

Other observations: _____

_____

_____

_____

_____

_____

RESOURCES

General Identification (ID) Resources

iNaturalist: Worldwide community science platform curated by experts. An independent nonprofit organization. inaturalist .org. Also an app for iPhone and Android.

Mushroom Observer: Community science platform dedicated to documenting and studying mushrooms, run by Nathan Wilson. mushroomobserver.org. Also an app for iPhone and Android.

Regional ID Resources

The Fungi of California: Overview of California mushrooms with photos, descriptions, and edibility information, curated by Michael Wood and Fred Stevens. mykoweb.com

Pictorial Key to Mushrooms of the Pacific Northwest: Pictorial key and overview of mushrooms found in the Pacific Northwest, curated by Danny Miller. alpental.com/psms/PNW Mushrooms/PictorialKey/index.htm

First Nature: Overview of mushrooms in the UK with photos, descriptions, and edibility information, curated by author Pat O'Reilly. first-nature.com/fungi

MycoKey Fungi of Temperate Europe: Phenomenal collection of very detailed infographic wheels on the mushrooms and fungi of temperate Europe, including spores. This PDF download features an excerpt from a book by Thomas Læssøe and Jens H. Petersen. mycokey.com/Downloads/FungiOf TemperateEurope_Wheels.pdf

The Hidden Forest: Overview of mushrooms in New Zealand with photos and edibility information, curated by Clive Shirley. fungi.co.nz

The Bolete Filter: Pictures and edibility information on North American boletes, curated by the Western Pennsylvania Mushroom Club. boletes.wpamushroomclub.org

General Education Resources

Mushroom Expert: Overview of mushrooms, with detailed scientific descriptions, but no edibility information, curated by author Michael Kuo. mushroomexpert.com

World of Fungi: Detailed encyclopedic information on the entirety of the fungal kingdom, curated by author David Moore. davidmoore.org.uk

Fungus Fact Friday: Blog with detailed posts on specific mushrooms and topics, curated by Thomas Roehl. fungusfactfriday.com

Cornell Mushroom Blog: Blog with detailed posts about various mushrooms and fungi, curated by Dr. Kathie T. Hodge. blog.mycology.cornell.edu

MycoGuide: Introduction to mushroom-forming fungi with great resources and links, curated by Patrick Leacock. mycoguide.com/guide/fungi

FungiKingdom: Defining the root meaning of mushroom names and morphology terms by Dianna Smith. fungikingdom.net

The Mycology Webpages: Broad overview of the fungal kingdom curated by David Malloch at the New Brunswick Museum. website.nbm-mnb.ca/mycologywebpages/Mycology WebPages.html

Community Science and Conservation Resources

Fungal Diversity Survey (FunDiS): Organization dedicated to documenting, sequencing, and conserving mushrooms in North America. fundis.org

Fungi Foundation: International organization dedicated to the discovery, documentation, and conservation of fungi, founded by Giuliana Furci. ffungi.org

SPUN: Coalition of scientists dedicated to studying the underground networks of mycelia worldwide, founded by Dr. Toby Kiers. spun.earth

Fungus Conservation Trust: Organization focused on the conservation of fungi in the UK. fungustrust.org.uk

Global Fungal Red List: Collection of threatened and endangered fungi from around the world. redlist.info

International Association for Lichenology: Organization dedicated to promoting the study and conservation of lichens. ial-lichenology.org

Fungal Taxonomy and Science Database Resources

MycoBank: Fungal taxonomy, nomenclature, DNA sequences, and species databank curated by International Mycology Association, Westerdijk Fungal Biodiversity Institute, and the German Mycological Society. mycobank.org

Index Fungorum: Taxonomic names and DNA sequences of fungi, curated by Kew Gardens, UK; Landcare Research, NZ; and Institute of Microbiology, China. indexfungorum.org /names/names.asp

Agaricales Database: Compendium of taxonomic relationships and names of gilled mushrooms, curated by Jacob Kalichman. agaric.us

Mushroom References: Curated list of scientific publications about mushrooms. mushroomreferences.com

International Commission on the Taxonomy of Fungi: Resources on the nomenclature, conventions, and naming of fungi. fungaltaxonomy.info

Mycology Collections Portal: A dynamic database that connects academic and community science mushroom observations worldwide. mycoportal.org

Mycopedia: Collection of fungal genomes for scientists organized by phylogenetic relationships. Includes databases of DNA sequences. mycopedia.org

UC Davis Wine Microbiology: Pictures and information on the yeasts and bacteria found in fermented beverages. wineserver.ucdavis.edu/industry-info/enology/wine -microbiology

Cultures & Cultivation Supply Resources

Commercial companies that sell mushrooms, grow kits, cultures, and cultivation supplies.

Far West Fungi: farwestfungi.com

Field & Forest Products: fieldforest.net

Fungi Perfecti LLC: fungi.com

North Spore: northspore.com

MycoSupply: mycosupply.com

Mushroom Mountain: mushroommountain.com

Phaff Yeast Culture Collection: Research collection of environmental and food-associated yeasts available for order, curated by the Boundy-Mills laboratory. phaffcollection.ucdavis.edu

Online Communities

Shroomery: Forum focused on the cultivation of entheogenic mushrooms. shroomery.org

Mycotopia: Forum discussions on cultivation and foraging. mycotopia.net

Fascinated By Fungi: Forum on Discord.

Mycology Clubs

North American Mycological Association (NAMA): A list of regional North American mycology clubs can be found on this site. namyco.org

Mycological Society of America (MSA): msafungi.org

Recommended Reading

21st Century Guidebook to Fungi, 2nd edition, by David Moore, Geoffrey D. Robson, and Anthony P. J. Trinci

Braiding Sweetgrass: Indigenous Wisdom, Scientific Knowledge and the Teachings of Plants by Robin Wall Kimmerer

Entangled Life: How Fungi Make Our Worlds, Change Our Minds & Shape Our Futures by Merlin Sheldrake

The Fifth Kingdom: An Introduction to Mycology, 4th edition, by Bryce Kendrick

FUNGI Magazine *The Hidden Kingdom of Fungi: Exploring the Microscopic World in Our Forests, Homes, and Bodies* by Keith Seifert

How to Change Your Mind: What the New Science of Psychedelics Teaches Us About Consciousness, Dying, Addiction, Depression, and Transcendence by Michael Pollan

The Lives of Fungi: A Natural History of Our Planet's Decomposers by Britt A. Bunyard

Molds, Mushrooms, and Medicines: Our Lifelong Relationship with Fungi by Nicholas P. Money

The Mushroom Hunter's Kitchen: Reimagining Comfort Food with a Chef Forager by Chad Hyatt

Mushrooms of the Redwood Coast by Noah Siegel and Christian Schwarz

Mycophilia: Revelations from the Weird World of Mushrooms by Eugenia Bone

Organic Mushroom Farming and Mycoremediation: Simple to Advanced and Experimental Techniques for Indoor and Outdoor Cultivation by Tradd Cotter

Radical Mycology: A Treatise on Seeing & Working with Fungi by Peter McCoy

AUTHOR'S NOTES

For further information,
please scan the QR code below.

TAXONOMIC TREE

NOTES

NOTES

NOTES

NOTES

NOTES

PHOTO CREDITS

Alan Rockefeller (IG: alan_rockefeller) Pg. 82, Ringless Honey Mushrooms (*Desarmillaria caespitosa*); Pg. 98, Yellow Morel (*Morchella americana*); Pg. 102, Eastern Caesar's Amanita (*Amanita jacksonii*); Pg. 104, Eastern Matsutake (*Tricholoma magnivelare*); Pg. 142, Dapperling (*Lepiota castaneidisca*); Pg. 192, Witches' Butter (*Tremella*); Pg. 194, *Cordyceps tenuipes*; Pg. 214, Spindles and Structured Lichens (*Cladonia/Cladina*); Pg. 215, Wolf Lichens (*Letharia*) **Ariel Rosen-Brown** Pg. 239, author photo **Christian Schwarz (IG: @biodiversiphile)** Pg. 126, Quinine Conk (*Laricifomes officinalis*); Pg. 144, *Cortinarius* sect. *Orellani* **Damon Tighe (IG: @DamonTighe)** Pg. 46, Short-Stalked Jack (*Suillus brevipes*); Pg. 68, Meadow Puffball (*Lycoperdon pratense*); Pg. 72, Wood Ears (*Auricularia*); Pg. 86, The Prince (*Agaricus augustus*); Pg. 90, Sawgills / Trainwreckers (*Neolentinus*); Pg. 98, Black Morels (*Morchella sextelata*), Woodchip Morel (*Morchella rufobrunnea*); Pg. 108, Short-Stemmed Russula (*Russula brevipes*); Pg. 110, Fried Chicken Mushrooms (*Lyophyllum decastes* group); Pg. 136, Destroying Angel (*Amanita bisporigera*); Pg. 146, Satan's bolete (*Rubroboletus eastwoodiae*) cross section; Pg. 152, Fool's Funnel (*Clitocybe rivulose*), Fool's funnel seen growing in grass; Pg. 154, Yellow staining agarics on grass; Pg. 158, Green-spored parasols (Chlorophyllum molybdites) on grass; Green-Spored Parasol (*Chlorophyllum molybdites*); Green-Spored Parasol (*Chlorophyllum molybdites*); Pg. 160 Earthballs (*Scleroderma*); Pg. 162, Cinnamon Bracket (*Hapalopilus rutilans*); Pg. 180, Bridal Veil Stinkhorn (*Phallus indusiatus*), Devil's Dipstick (*Mutinus elegans*); Pg. 184, Woolly Bird's Nest Fungus (*Nidula niveotomentosa*); Pg. 186, Close-up of cannonball or peridiole; Pg. 194, *Ophiocordyceps dipterigena*; Pg. 198, Rust pycnia (*Uromyces aureus*); Pg. 200, Smut (*Ustilago*), Brome Smut Fungus (*Ustilago bullata*); Pg. 214, Beard Lichens (*Usnea*); Pg. 215, Map Lichens (*Rhizocarpon*) **Danny Newman (IG: @kallempero)** Pg. 200, Huitlacoche, or Corn Smut (*Mycosarcoma maydis*) **Mandie Quark (IG: @mandie_quark)** Pg. 33, *Polyporus umbellatus*; Pg. 122, The chaga fruiting body is a toothy ragged crust. **Noah Siegel (IG: @mychobo)** Pg. 180, Devil's Fingers (*Clathrus archeri*) **Taye Bright (IG: @symbiiotica)** Pg. 86, Horse Mushroom (*Agaricus arvensis*) **Teresa Mycelia (IG:@teresa. mycelia)** Pg. 198, Cedar Apple Rust (*Gymnosporangium juniperi-virginianae*) **Warren Cardimona (IG:@warren.cardimona)** Pg. 144, Deadly Webcaps (*Cortinarius* sect. *Orellani*)

Dr. Gordon Walker is an award-winning speaker,
science communicator, published author, and
social media star (@FascinatedByFungi) with
more than three million cumulative followers.
He has a PhD in biochemistry and molecular
biology from UC Davis, where his research
focused on the microbial ecology of wine
fermentations, bioprocess control, and
yeast genetics. He is a passionate advocate
for mushrooms, fungal conservation, and
environmental sustainability.

This journal is intended to help identify mushrooms, but any readers intending to eat wild mushrooms should get confirmation from multiple other sources. Being able to accurately identify mushroom species is essential for foraging, often requiring years of study and hands-on experience. There are toxic look-alike mushrooms that can be confused for edible varieties. The author and publisher make no representations regarding the safety of foraging or consuming wild mushrooms and specifically disclaim any responsibility for any misidentification of mushroom species or health problems or loss resulting from the ingestion of mushrooms or the use of any information contained in this book. Persons who consume mushrooms or other potentially dangerous fungi do so at their own peril and risk illness and even death. Consuming mushrooms of unknown origin should be avoided, and even some mushrooms that are nontoxic for most people may make some people ill.

CLARKSON POTTER/PUBLISHERS
An imprint of the Crown Publishing Group
A division of Penguin Random House LLC
1745 Broadway
New York, NY 10019
clarksonpotter.com
penguinrandomhouse.com

A Clarkson Potter/Publishers Trade Paperback Original

Adapted from *Dr. Fun Guy's Passport to Kingdom Fungi* by Dr. Gordon Walker (New York: Ten Speed Press, 2025).

Photo credits are located on page 238.

Editor: Harry Tunggal
Designer: Annalisa Sheldahl | Art director: Danielle Deschenes
Production editor: Abby Oladipo
Production: Kelli Tokos
Compositor: Barbara Peragine
Copyeditor: Ethan Campbell | Proofreaders: Sarah Rutledge and Robin Slutzky
Marketer: Chloe Aryeh

Manufactured in Malaysia

10 9 8 7 6 5 4 3 2 1

The authorized representative in the EU for product safety and compliance is Penguin Random House Ireland, Morrison Chambers, 32 Nassau Street, Dublin D02 YH68, Ireland, https://eu-contact.penguin.ie.